SPACE, TIME, AND PRESENCE IN THE ICON

Ashgate Studies in Theology, Imagination and the Arts

Series Editors

Trevor Hart, St Mary's College, University of St Andrews, Scotland
Jeremy Begbie, Duke University, USA
Roger Lundin, Wheaton College, USA

What have imagination and the arts to do with theology? For much of the modern era, the answer has been, 'not much'. It is precisely this deficit that this series seeks to redress. For, whatever role they have or have not been granted in the theological disciplines, imagination and the arts are undeniably bound up with how we as human beings think, learn and communicate, engage with and respond to our physical and social environments and, in particular, our awareness and experience of that which transcends our own creatureliness. The arts are playing an increasingly significant role in the way people come to terms with the world; at the same time, artists of many disciplines are showing a willingness to engage with religious or theological themes. A spate of publications and courses in many educational institutions has already established this field as one of fast-growing concern.

This series taps into a burgeoning intellectual concern on both sides of the Atlantic and beyond. The peculiar inter-disciplinarity of theology, and the growing interest in imagination and the arts in many different fields of human concern, afford the opportunity for a series which has its roots sunk in varied and diverse intellectual soils, while focused around a coherent theological question: How are imagination and the arts involved in the shaping and reshaping of our humanity as part of the creative and redemptive purposes of God, and what roles do they perform in the theological enterprise?

Many projects within the series have particular links to the work of the Institute for Theology Imagination and the Arts in the University of St Andrews, and to the Theology Through the Arts programme in Cambridge.

Other titles in the series

Theological Aesthetics after von Balthasar
Edited by Oleg V. Bychkov and James Fodor

Faithful Performances
Enacting Christian Tradition
Edited by Trevor A. Hart and Steven R. Guthrie

Space, Time, and Presence in the Icon

Seeing the World with the Eyes of God

CLEMENA ANTONOVA
Royal Flemish Academy of Belgium for Science and the Arts

with a preface by
MARTIN KEMP

ASHGATE

Published by
Ashgate Publishing Limited
Wey Court East
Union Road
Farnham
Surrey, GU9 7PT
England

Ashgate Publishing Company
Suite 420
101 Cherry Street
Burlington
VT 05401-4405
USA

www.ashgate.com

British Library Cataloguing in Publication Data
Antonova, Clemena, 1970–
Space, Time, and Presence in the Icon: Seeing the World with the eyes of God.
– (Ashgate Studies in Theology, Imagination and the Arts)
1. Space and time in art. 2. Space and time – Religious aspects – Christianity.
3. Icons, Byzantine. 4. God (Christianity) – Eternity. 5. Florenskii, P. A.
(Pavel Aleksandrovich), 1882–1937. I. Title II. Series
246.5’3–dc22

Library of Congress Cataloging-in-Publication Data
Antonova, Clemena, 1970-
Space,Time, and Presence in the Icon: Seeing the World with the Eyes of God /
Clemena Antonova.
p. cm. – (Ashgate Studies in Theology, Imagination, and the Arts)
Includes bibliographical references and index.
1. Space and time in art. 2. Perspective. 3. Icons. I Title.
N7430.7.A58 2009
246’.53–dc22 2009017751

ISBN 9780754667988 (hbk)
ISBN 9780754697381 (ebk)

Mixed Sources
Product group from well-managed forests and other controlled sources
www.fsc.org Cert no. SA-COC-1565
© 1996 Forest Stewardship Council
FSC

Printed and bound in Great Britain by
MPG Books Group, UK

Contents

To my parents and Jean
with love and gratitude

List of Illustrations

Acknowledgements

The present book represents a shorter, revised version of my D.Phil. thesis in Art History at Oxford University, which I defended in 2006. I therefore would like to express my deepest gratitude to my thesis supervisors, Prof. Martin Kemp and Prof. Paul Crowther, for their unfailing help. I consider myself truly fortunate in having been able to work with Prof. Kemp and Prof. Crowther and I would like to thank them for their patience, the long hours they spent on my thesis, as well as for everything I have learned from them in the process of work.

Over the years I have consulted a number of scholars on various aspects of my work. My thesis examiners, Prof. Simon Franklin and Prof. James Elkins, encouraged me in my book project and gave me valuable guidance, which proved extremely useful in the process of revising my work. Further, I am particularly grateful to Prof. Richard Swinburne, Dr. Joseph Munitiz, Prof. Moshe Barasch, Dr. Jeremy Begbie, Dr. Ivan Christov and Rev. Hugh Wybrew for helping me with the theologically-related material of my text. Prof. Dimiter Zashev, Prof. Edward Craig, Dr. William Newton-Smith and Prof. Christopher Janaway helped me with passages on various philosophical issues. For the Russian material, I could consult Prof. Pamela Davidson, Prof. Franklin, Prof. David Shepherd, and Dr. Andrei Shishkin, while Prof. Henry Maguire, Prof. Cyril Mango, and Prof. Robin Cormack were kind to answer my questions relating to Byzantine art.

My thanks are also due to Somerville College, Oxford, not only for the scholarship they granted me, but also for making me feel welcome during my stay in Oxford. My work was also made possible by the generous financial support of the Pirie Reid Fund and the gracious help of Anton Tchitchekliev. I am also grateful to St. Deiniol's Library in Northern Wales for the residential scholarship, which allowed me to use the theology resources at the library. Further, I had ideal research conditions during my Mellon Fellowship at the Institute for Advanced Studies in the Humanities in Edinburgh, which allowed me to work on my book project. A lot of the reading was done while I was teaching for the Francis E. Kelly Summer Programme at Oxford, directed by Prof. Basil Valente, a colleague and a dear friend.

I am grateful to the Scouloudi Foundation in association with the Institute of Historical Research London for giving me a Historical Award.

Also, I have to mention my appreciation of the exemplary professionalism of my editors at Ashgate, Kirsten Weissenberg and Sarah Lloyd.

Finally, my deepest gratitude is due to my mother Maria Raicheva, my father Bojidar Antonov, my closest friend Vera Genova and my husband Jean Crombois. They have all been amazingly supportive. I also thank my friends at Oxford –

Theodore, Ivaylo, Stelian, Nora and Nelly – for their intelligent company and constant help.

Permissions

Some parts of the following text have been published previously:

'"Beauty Will Save the World": The Revival of Romantic Theories of the Symbol in Pavel Florensky's Works', *Slavonica,* 14/1 (2008): 44–56

with Martin Kemp, '"Reverse Perspective': Historical Fallacies and an Alternative View' in Michele Emmer, *The Visual Mind II,* (Cambridge, MA: 2005), pp.399–433

'Florenky's "Reverse Time" and Bakhtin's "Chronotope" – a Russian Contribution to the Theory of the Visual Arts', *Slovo,* 15/2 (2003): 101–114

Preface

Eastern Orthodox icons have been largely defined in Western art history by what they are not – non-naturalistic, non-perspectival and non-progressive. While Western art is conventionally seen as striding forward into the Renaissance revolution and ultimately into the perceptual subtleties of Impressionism, icons remained stuck in an immobile mode, bereft of the earlier naturalism of ancient Greek and Roman art, and unable to absorb what progressive artists in the West were doing. Even for authors who greatly admired icons, the definition of their visual qualities invariably took Western art as an implicit foil.

Russian scholarship in the 20th century started from this point, but inverted it, by arguing that to be non-naturalistic and non-perspectival was to be more "true" to seeing than the discredited conventions of Western naturalism. There was a specifically nationalistic edge to this argument. Russian icons had, in effect arrived at the progressive solution of the Cubists' vision of form and space centuries before Picasso and Braque. "Reverse perspective" was a creative denial of linear perspective, positing an alternative optics. There is, however, something paradoxical about "reversing" something that came later.

What is needed is a fresh start. We need to ask what icons actually do, at the highest level of spiritual and liturgical functionality in their own right, rather than seeing them as in opposition or in counterpoint to something. This new questioning is what Clemena Antonova is doing, both providing a fresh start and taking us some way along roads that are clearly heading in the right direction.

We could say in a clichéd sense that icons are "timeless", but if we take that idea not as a cliché but as something that is profoundly embedded in the function of the sacred image, we might make some progress into understanding why the images are "spaceless" in terms of the measured optical spaces of a perspective painting. Icons purvey spiritual images of eternal verity, outside the time and space of the temporal viewer in the here and now. The best the transient viewer on earth can hope to do is to glimpse a fragment of the eternity that awaits the souls of the redeemed.

We might also say in a clichéd sense that icons, particularly those of Christ, the Virgin and single saints, have a great sense of "presence". This is achieved through the direct and simple presentation of the sacred figures, often in austerely frontal modes. But if we take "presence" in a more theological sense, seeing the sacred persona as in some way embodied in their image, we can begin to gain a better idea of the role of the icon for the viewer. The nature of that presence is a subtle matter, as Antonova recognises.

The quality of being "outside time" and the sense of "presence" to which the images aspire explains why the types of representation do not undergo revolutionary transformations. There are no Masaccios or Caravaggios in icon art. A very good witness to what an Orthodox worshipper expects of an image is Gregory Melissenos, a representative of the court of the Holy Roman Emperor in Constantinople at the Council of Ferrara and Florence in 1438-9. He was unsettled by what European artists had been doing in the 14th and 15th centuries. He reported that "when I enter a Latin church, I can pray to none of the saints depicted there because I recognise none of them. Although I do recognise Christ, I cannot even pray to him because I do not recognise the manner in which he is depicted". The principles of timelessness and presence had been violated, and the particularities of realistic space and form were responsible.

The issues of form and function that Antonova's analysis of "reverse perspective" illuminates go beyond the immediate issues of the portrayal of space. They go to the very heart of the matter of how Icons have consisitenlty worked for their users over the centuries.

Martin Kemp
August 2009

Introduction

The writing of this book is motivated by a number of questions which have baffled viewers of Christian painting for long. The first – and most important – concerns whether there is any essential feature of *icon art* that is not confined to a particular period but runs throughout its history. In other words, what makes a visual image an *icon*? What grounds our tendency to speak and think of certain works as "icons" rather than simple "pictures" or "paintings"? Put more specifically, what does, say, a sixth-century image from the Mount Sinai share in common with Andrey Rublev's early fifteenth-century *Holy Trinity* that enables us to describe both as "icons"?

Such questions can be described as essentialist. The author is fully aware of the inherent dangers of such an approach, but considers it worthwhile, nevertheless, to address the icon in terms that transcend concrete, narrowly defined historical periods and geographical locations. After all, the icon as a form is still very much alive even though we associate it mainly with the medieval period. Further, whilst icons may have been the art *par excellence* of the Byzantine Empire, they were produced extensively outside the realm of Byzantium as well, and include non-Orthodox audiences and non-Orthodox artists.

Given that the icon is not tied rigidly to highly localized historical, geographical or religious contexts, I will identify a decisive factor (there may be others, too, of course) that justifies a unified conception of icon art. It is the pronounced persistence of a certain principle concerning the organization of pictorial space and, by implication, of time in a large group of paintings over a long period. This spatial formula has been commonly referred to as "reverse perspective".

Whatever "reverse perspective" means, it is clearly a mode of representation which is relatively unusual for a Western viewer accustomed to images constructed according to the laws of standard, linear pictorial space. This is the main reason why icons are frequently described as "distorted". The present work, therefore, has a practical end of providing a Western audience with a guide to reading icons in space. This is the underlying concern of the discussion of "reverse perspective". The analysis of Rublev's famous icon at the end is meant to serve as an illustration of the way that the categories proposed in this book work in concrete, visual terms.

Two problems arise at this stage. One has to do with the lack of a convincing theory of "reverse perspective". The view proposed here relies on making a connection between "reverse perspective" and the theological dogma of timeless eternity. In this way, another problematic issue makes itself felt, namely the temporal dimension of pictorial time in general. These two problems are intertwined, and the first part of the book will explore them in detail. Thus, Chapter 1 looks at the problem of time in the visual arts in general and the icon in particular. The controversial nature of this issue is signalled by paying attention to a well-established tradition in art criticism which denies temporality to the visual arts. Chapter 2 considers some of the major writings on "reverse perspective", mainly by Russian authors, some of which are little known in the West. The generally accepted idea, reflected in the terminology itself, that "reverse perspective" turns around or reverses the laws of linear perspective is challenged on several grounds. This is why, while we stay with the term "reverse perspective" as it has gained currency, we will keep it in inverted comas to signal that this terminology is unsatisfactory and that eventually a new one should be developed.

Before proceeding to an elaboration of an alternative view, it is necessary to address, in Chapter 3, the question of the importance and function of "reverse perspective". The principle of constructing space in icons has staunchly persisted, even in the face of the triumphal advent of mathematical space, over a long period of time. Probably not all, but an overwhelming number of icons have been done in "reverse perspective". That kind of continuity makes it reasonable to assume that this principle of pictorial space has a significance that has made it endure amid other changes of style. It has been suggested that "reverse perspective" is an element of the canon of the art of the icon and it serves as a "container" of presence. In other words, the spatial construction has the function of a guarantee of the presence of the prototype in the image.

The last and longest chapter, Chapter 4, plays on all three themes of time, space and presence. It proposes an explanation of the construction of pictorial space in the icon by drawing a structural analogy with the theological dogma of timeless eternity. The main principle of "reverse perspective" is understood not as the turning around of the laws of linear perspective (as in the widely accepted view), but as the representation of the "simultaneous planes" of an object, i.e., planes which cannot be seen from a fixed position at one moment of time. The viewer's perception of an icon, defined in these terms, can be compared to the "vision" of a God, who exists beyond time and to whom, therefore, all aspects of the objects in the world would appear at once, simultaneously. In other words, such a divine being would have no point of view in space. The transcendental nature of icons, frequently remarked upon but almost never convincingly explained, is interpreted here as founded on a spatial construction which allows the beholder to experience a form of divine vision which transcends the human constraints of space and time. Some attention is given to the nature of the connection between the theological, conceptual notion and the artistic practice, expressible in visual terms. The possibility of a "theology through the arts" is seen as both promising and problematic.

Whilst the basic impetus of my book concerns the conceptual basis of the icon, I am also aware of the importance of historical conditions under which key notions concerning the icon have developed. Thus, a second aspect of my approach to the material in the book has to do with tracing the role of interpretative traditions associated with icon art, for example, Antique and Late Antique pagan and Christian Neoplatonism, Byzantine theology of the image of the eighth and ninth centuries, German romantic philosophy at the end of the eighteenth century, and Russian religious philosophy at the beginning of the twentieth century as represented mainly by Pavel Florensky. In some ways, the latter closes the circle as it was strongly influenced by all the other traditions, but, interestingly, it also points ahead to tendencies we usually associate with post-modernism (section on Florensky and Heidegger in Chapter 3). Thus, for example, the problem of presence, which is of fundamental importance to this book, was first formulated by Neoplatonists as Proclus and Plotinus. It was then borrowed and elaborated on by writers within the Byzantine tradition as St. John of Damascus and St. Theodore the Studite and applied, in a Christian guise, to the theology of the icon. The problem was revisited by the German romantics as Schelling and Schlegel, who, in turn, had a decisive impact on Russian thought. Florensky's concern with presence in the holy image goes back, largely via romantic philosophy, to Neoplatonism and Byzantine theology, but it also sounds remarkably close to the late Heidegger. This outline, of course, simplifies matters but it also indicates a major thrust of human thought which underlies the main connotation of a term or concept.

A third concern underlying the book as a whole is to combine Western scholarship with work done in Russia. The revived interest in the icon in Russia goes back to the middle and the second part of the nineteenth century and forms an important part of Russian *fin-de-siecle* intellectual history. Very few of the Russian writings on the subject are actually familiar to the Western reader. One of the intentions of the present book is to make at least a fraction of these works better known. The focus has fallen on one of the most interesting figures from the first part of the twentieth century – Pavel Florensky (1882–1937), priest, religious philosopher and scientist, who combined all these fields of expertise in his study of the Russian icon. Florensky has been considered of importance not so much for having provided solutions to various problems under our attention, but for having asked some of the questions we will be concerned with. For instance, he is the only writer I am aware of who attempted to offer a systematic explanation of the temporal dimension of the icon (section in Chapter 1), he wrote the most influential essay on "reverse perspective" (section in Chapter 2), and was consistently interested in the problem of presence in the image and its implications for modern culture (section in Chapter 3). My own approach to Florensky has been critical, but at the same time the Russian author has proved inspirational in many ways. The concept of "simultaneous planes", which is decisive for formulating the major thesis of the present work, is an elaboration of Florensky's own term of "supplementary planes" (section in Chapter 4). Florensky's writings on the icon are seen in the context of intellectual developments in Western Europe at the time (Cubism, the revival of Occultism, etc.), but also in their embeddedness in Russian intellectual history (work done at the Russian Academy of Artistic Sciences in Moscow, writings by Bakhtin, Viacheslav Ivanov, Zhegin, Boris Uspensky, etc.).

It is only in recent years that the need has been felt to bring together the contributions of Russian and Western thinkers. The attempt to do so can produce fascinating results. Not least, Russian thought has frequently put the emphasis on distinct aspects of basically the same problems as have interested Western scholars and thus has illuminated these problems from different, unexpected perspectives. The opposite is, of course, true as well – the Russian approach has much to gain by an insight into Western interpretations. That the two traditions have been kept largely separate is, in fact, to the loss of both. Florensky himself is a very good example of the exciting insights that can be achieved on the basis of an in-depth familiarity with Western, especially German scholarship in tackling typically Russian intellectual themes.

Ultimately, then, the material covered by this book and the approach towards it are part of a project that aims to provide the modern viewer with an understanding of the icon. There are numerous definitions of the term in dictionaries, encyclopaedias and books. The following work does not intend to supplant them, but to make more concrete the frequently vague claims about the spirituality, holiness, transcendental nature of icons and to add a further dimension to the theological meaning of sacred images. It is my belief that the principle of pictorial space of the icon should not be an exclusively art historical issue, but a necessary element of the understanding of the icon and should, therefore, be included in its definition. The outline of "icon" in the glossary of terms is an attempt to work out such a definition.

Chapter 1
The Role of Time in the Pictorial Art

In the present chapter, it is argued that time is a fundamental organizing principle of pictorial art. The widely accepted, but not unchallenged, distinction between spatial and temporal arts, popularized by Lessing, is considered to be flawed, since it is unable to account for the complex nature of the arts. Further, the spatial organization that underlies the image is accepted to entail, by necessity, the conception of time. Before considering the concrete case of the temporal dimension of the case of the icon, as based on the writings of Pavel Florensky, some attention needs to be devoted to the underlying assumptions of the widely accepted space-time categorization of the arts.

The Problem of Time in the Visual Arts – Seeing a Picture "in the Twinkling of an Eye"

Visual art is usually taken to be a matter of the manipulation of the material in space while the temporal dimension is often disregarded. This has largely been so both on the intuitive level and in the realm of academic discourse. Indeed, most would agree with W. Thomas Mitchell when he says that "nothing [...] seems more intuitively obvious that the claim that literature is an art of time, painting an art of space".[1] It seems evident that painting has "no natural temporal extension".[2] In the critical history of the visual arts the problem of time has remained peripheral and when touched upon, the prevailing view has tended to suppress the temporal dimension. Otto Pächt, for instance, is expressing a wide-spread view, when he describes the history of pictorial art as "a series of repeated attempts to smuggle the time factor into a medium which, by definition, lacks the dimension of time".[3]

---

[1] W. J. Thomas Mitchell, *Iconology: Image, Text, Ideology*, (Chicago and London, 1986), p. 95.

[2] D. Ades, 'Art and Time in the Twentieth Century' in Kristen Lippincott, Umberto Eco, Ernst Gombrich et al, *The Story of Time*, (London, 1999), p. 202.

[3] Otto Pächt, *The Rise of Pictorial Narrative in Twelfth Century England*, (Oxford, 1962), p. 1. Pächt's position is more complicated, as, for instance, in a later work he devotes a section to the problem of "unfamiliar notions of time" in the visual arts (Otto Pächt, *Practice of Art History: Reflections on Method*, (London, 1999, first in German in 1986), pp. 41–5).

What Mitchell calls "the tradition of denying temporality in the visual arts"[4] can be attributed, to a large extent, to the impact of Lessing's *Laocoön* (1766). The division of the arts into arts of time and those of space is traditional and had long been in existence before the publication of the *Laocoön.* Gombrich points out that even Joseph Spence and the Comte de Caylus, whom Lessing cited as having ignored the space-time distinction, actually are explicitly aware of it.[5] Most famously, Leonardo maintained that "painting immediately presents to you the demonstrations its maker has intended", while "the works of the poets must be read over a long span of time" and further that painting "simultaneously conveys the proportional harmony of which the parts of the whole are composed", while poetry describes "the configurations of particular objects more slowly than is accomplished by the eye".[6] Leonardo's position, however, is a bit more complicated, I think, but I will return to that later.

It was Lessing (1729–1781), however, who was the first to systematically treat this question and popularize this distinction. His position has been hugely influential since and was questioned only at the beginning of the twentieth century. Goethe (1749–1832) bears witness to its impact at the time:

> "One would have to be a young man again to realize the effect wrought upon us by Lessing's *Laokoon* [...] transported us from the region of slavish observation into the free fields of speculative thought. The long misunderstood *ut pictura poesis* was at once set aside. The difference between picture and poetry was made clear – the peaks of both appeared separate, however near might be their bases."[7]

In his 1920s 'Creative Credo' Paul Klee's (1879–1940) response is quite different:

> "In Lessing's *Laokoon*, on which we squandered study time when we were young, much fuss is made about the difference between temporal and spatial art. Yet, looking into the matter more closely, we find all this is but a scholastic delusion. For space, too, is a temporal concept."[8]

---

4 Mitchell, *Iconology: Image, Text, Ideology*, p. 99.

5 Ernst Gombrich, 'Lessing (Lecture on a Master Mind)', *The Proceedings of the British Academy*, (London, 1957): 139. Precedents for Lessing's distinction are also discussed by Nikolas Schweizer in his book *The Ut Pictura Poesis Controversy in Eighteenth Century England and Germany*, (Bern, 1972).

6 Martin Kemp, (ed.), *Leonardo on Painting*, (New York and London, 1989), p. 23.

7 Goethe, *Poetry and Truth: From My Own Life*, vol. 2, (London, 1908), p. 282.

8 Paul Klee, 'Creative Credo', in Herschel Chipp, (ed.), *Theories of Modern Art: A Source Book by Artists and Critics*, (Berkeley and London, 1968), p. 184.

Apparently it seemed to Goethe and others at the time that Lessing had solved a genuine problem. But had he, later generations would ask? If there is a fundamental difference between painting and poetry, between image and language, would it come down to a space-time distinction? Besides, in Goethe's own metaphor it would appear that things are very much a matter of emphasis, depending on which part of the intersection between the two figures of painting and poetry we would choose to investigate. The closer to the "bases" is our analysis the more likely it would be that we emphasize the unity of the two arts. The nearer to the "peaks" we go the more we will tend to see differences and specificities. William Wimsatt in his essay on Lessing is well aware of that, while he himself tends to the position that "the arts are in fact considerably different from one another".[9] Svetlana and Paul Alpers, similarly, see the statement, attributed to Simonides (556–468 BC),[10] that poetry is a speaking picture and painting a mute poesy as "more witty than truthful"[11] and put the stress on the profound difference among the arts in their conditions and "languages".[12] Gombrich has a like concern to show that "the means of visual art cannot match the statement function of language".[13]

On the other hand, Horace's tradition of the *ut pictura poesis* (literally "as poetry, so painting") belongs to another stream of literary and art criticism in its emphasis on the unity of the arts, which has been a constant theme of aesthetics as an academic discipline. It was brought to an extreme in the seventeenth and eighteenth centuries and especially by French and English critics, as Rensslelaer Lee points out in a well-known article. Lee is right to find Dryden's comparison between the two arts in 'A Parallel of Poetry and Painting' (1695) "absurdly elaborate",[14] while fifty years after Dryden the Abbé Batteaux published an essay in the same spirit under the telling title 'The Fine Arts Reduced to a Single Principle' (1746).[15] In the second half of the nineteenth century Walter Pater bears witness to a similar trend when he says at the opening of his essay 'The School of Giorgione' (1877): "It is a mistake of much popular criticism to regard poetry, music, and painting

---

[9] William Wimsatt, 'Laokoon: An Oracle Reconsulted' in his *The Day of the Leopards: Essays in Defense of Poetry*, (New Haven and London, 1976), p. 4.

[10] Simonides' doctrine is mainly familiar through by Plutarch, who quotes Simonides in the *Moralia*.

[11] Svetlana Alpers and Paul Alpers, '*Ut Pictura Noesis*? Criticism in Literary Studies and Art History', *New Literary History*, 3/3, (1976): 446.

[12] Ibid., 458.

[13] Ernst Gombrich, 'The Evidence of Images', in Charles Singleton, (ed.), *Interpretation: Theory and Practice*, (Baltimore, 1969), p. 97. See also by the same author the review of Charles Morris's 'Signs, Language and Behaviour', *Art Bulletin*, 31 (1949): 72, as well as Ernst Gombrich, *Art and Illusion*, (London, New York, 1960).

[14] Rensslelaer Lee, '*Ut Pictura Poesis*: The Humanistic Theory of Painting', *Art Bulletin*, 22 (1940): 202.

[15] Abbé Batteux, 'The Fine Arts Reduced to a Single Principle', in Susan Feagin and Patrick Maynard (eds.), *Aesthetics*, (Oxford and New York, 1997), pp. 102–105.

– all the various products of art – as but translations into different languages of one and the same fixed quantity of imaginative thought".[16] Irving Babbitt provides a useful insight into the state of the confusion of the arts which had been reached at the time in his work *The New Laocoon* (1910). The analogies between painting and poetry that Aristotle in the *Poetics* and Horace in the *Ars poetica* had drawn had grown into virtual identifications, something which the ancient texts had never intended.[17] In Jean Hagstrum's words, Aristotle's view on the arts would make them appear as cousins rather than sisters,[18] while Horace's dictum should not be interpreted as "let a poem be like a painting", but rather "as a painting, so also a poem" or "as sometimes in a painting, so occasionally in poetry".[19]

The Antique analogies between the arts could just as well serve as an authority to the other stream of thought which attempts to categorize the arts according to various criteria. There have been different systems of categorization and the division between spatial and temporal arts is one among the many. The question is if it is a viable one and if it discloses anything of universal significance about the nature of the arts. If we come to feel that Lessing has overstrained the divisions, this too should be understood in a historical context. Lessing was reacting against the other extreme just mentioned. It has been noticed how he was anticipated by La Fontaine (1621–1695), who, ironizing the tradition of *ut pictura poesis* at his time, had remarked:

> Les mots et les couleurs ne sont choses pareilles
> Ni les yeux ne sont les oreilles.[20]
>
> (Words and colours are not comparable things
> The eyes are not the ears).

The problem of time arose in the context of debates on the nature of the visual and/versus the verbal arts. Etienne Sourian points out in his essay "Time in the Plastic Arts' (1945) that the visual arts, as all arts in general, involve a "*psychological* time of contemplation".[21] This is valid not only in the cases of architecture and sculpture in the round, where a moving spectator is obviously

[16] Walter Pater, 'The School of Giorgione' (1877) in his *The Renaissance*, (Oxford and New York, 1986, first in 1873), p. 83.

[17] On that, see Lee, '*Ut Pictura Poesis*: The Humanistic Theory of Painting', especially 197–203.

[18] Jean Hagstrum, *The Sister Arts: The Tradition of Literary Pictorialism and English Poetry from Dryden to Gray*, (Chicago, 1958), p. 6.

[19] Ibid., p. 9.

[20] I am quoting from Lee, '*Ut Pictura Poesis*: The Humanistic Theory of Painting', 203.

[21] Etienne Sourian, 'Time in the Plastic Arts', *Journal of Aesthetics and Art Criticism*, 7/4 (1949): 295.

presupposed. Sourian rightly maintains that even in the case of painting a "time of contemplation is required".[22] John Dryden, too, realized this aspect of the problem and he says in his essay "A Parallel of Poetry and Painting" that, while the elements of a picture "are to be discerned all at once, in the twinkling of an eye", this is so only provisionally "if sight could travel over so many objects all at once, or the mind could digest them all at the same instant or point of time".[23] The author gives us an example with Poussin's *The Institution of the Blessed Sacrament*, where the many figures and the various actions they perform would require that the picture be "seen by intervals" and "considered at leisure".[24] I do not think Leonardo would have disagreed with this observation as he was talking about something else – a picture can convey a total impression at first glance, whatever other characteristics a further contemplation might reveal, in a way that poetry cannot. Therefore, the perception of harmony depends on the medium of presentation. In poetry, the effect of harmony is, as if disrupted, since "the words with which he [the poet] delineates the elements of beauty are separated from one another by time, which leaves voids between them and dismembers the proportions".[25] On the other hand, the effect of painting is total in that "it simultaneously conveys the proportional harmony of which the parts of the whole are composed".[26]

The problem posed here would centre round the connotation that is assigned to the concept of simultaneity. A modern writer, like Jonas Hans, uses the term "simultaneous" very much in the sense of Leonardo. Sight, according to Hans, is a sense of simultaneity, as it is capable of encompassing a visual field at one moment. Consequently, Hans believes, that sight is intrinsically less temporal than the other senses. The "nobility of sight" is said to be due to the fact that "the very contrast between eternity and temporality rests upon an 'idealization' of the 'present' experienced visually as the holder of stable contents as against the fleeting succession of non-visual sensations".[27] Leonardo's argument, I believe, was driving in the same direction.

Lew Andrews makes an important observation, when he draws attention to the importance of the comparison which Leonardo makes between painting and music. Music is the sister of painting, Leonardo says and it creates "harmony through the conjunction of proportional parts during the same span of time" (Cod. Urb.16–16v).[28] It is the last phrase – *nel medesimo tempo* – that we are interested in, as it refers to Leonardo's understanding of simultaneity. As Andrews points out,

---

[22] Ibid.

[23] John Dryden, *The Essays of John Dryden*, ed. W. Ker, (New York, 1961), p. 131.

[24] *op. cit.*, p. 132.

[25] Kemp (ed.), *Leonardo on Painting*, p. 24.

[26] Ibid., p. 23.

[27] Hans Jonas, *The Phenomenon of Life: Toward a Philosophical Biology*, (Chicago and London, 1982, first in 1966), p. 145.

[28] I am citing from Lew Andrews, *Story and Space in Renaissance Art: The Rebirth of Continuous Narrative*, (Cambridge, 1999), p. 64; Andrews' translation.

a *tempus* in Renaissance musical terminology is a unit of musical time of relatively brief but definite duration and thus Leonardo is "speaking in all likelihood, of a span of time".[29] Andrews draws the conclusion that "by drawing music into his argument, Leonardo ultimately brings out the successive character of painting rather than instantaneous aspects of either art".[30] Surely, Leonardo's intention could hardly have been that. The idea remains, however, that Andrews draws attention to Leonardo's view that the perception of painting requires time, even if it is a relatively short period.

This position, however, gives rise to the following suspicion – is it not too simplistic to attribute different kinds of perception, required by the different media of painting and literature to a quantitative difference in the time it takes us to apprehend each? And, strictly speaking, does not the "twinkling of an eye" take place in time too, as "instantaneous experience is an impossibility, biologically and psychologically".[31]

To analyze the structure of the work in terms of the conception of time it conveys presupposes at the same time the ability to distinguish between *types* of pictorial time. As was already mentioned, Sourian talks of "psychological time", involved in the act of perception, an idea noticed by earlier writers. Much more interestingly, he draws attention to another temporal mode in art which he calls "*intrinsic* time".[32] No further discussion is offered and it has actually been noticed, and with good reason, that Sourian's analyzes of the illustrations he uses fail to show what exactly constitutes "inherent time".[33] Still, Sourian's article seems important as it points in a direction that might get us to fruitful results. The crux of the matter, I believe, lies exactly in exploring "the artistic time inherent in the texture itself of a picture or a statue, in their composition, in their artistic arrangement"[34].

In the attempt to differentiate between types of time in pictorial art it would be useful to distinguish between the following three types. Firstly, we can talk of *external* (to the picture), scanning time or time spent by the viewer in perceiving the image, which could vary in length. This refers to what Martin Kemp calls "the *externally imposed* context within which the viewing takes place".[35] According to Kemp, this "would involve both prescribed dimensions, as when an image is used within an orchestrated ritual, as in church liturgy, and a voluntary component, when viewing is commenced and terminated, according to the viewer's choice, as when perambulating in an art gallery or church. There is a sense of the time that

---

29 Ibid., p. 65.

30 Ibid., p. 64.

31 John Dewey, *Art as Experience*, (London, 1934), p. 220.

32 Sourian, 'Time in the Plastic Arts', 297.

33 G. Giovannini, 'The Method of Study of Literature in Its Relation to the Other Fine Arts', *Journal of Aesthetics and Art Criticism*, 8 (1949): 191.

34 Sourian, 'Time in the Plastic Arts', p. 297.

35 Communication from Martin Kemp, April 2003.

the viewer 'ought' to look at the image, whether for aesthetic or other purposes. Praying to an icon, or lighting a candle in front of one, are acts that carry the implication that a certain amount of time is spent absorbing the image through one's eyes".[36]

A completely different temporal dimension is the *internal* (to the picture), "perspectival" time. It is generated by the spatial organization of the image and depends on the idea that specific treatments of images in space express specific conceptions of time, as happens most notably in narrative. A third kind is the internal-external pictorial time, which relies on the viewer's reading of iconographical and narrative clues, provided by the image. In relation to this type of pictorial time, Kemp continues, "some images, particularly complex ones, such as William Frith's elaborate telling of multiple stories in his *Railway Station*,[37] demand a sustained interplay between the time of viewing and the time-frame of the internal stories if the content is to be 'read' at the required level. Paintings with sets of symbolic allusions also demand extended time for viewing, but this time is not linked to overt narrative content within the image (as in the picture of the standing saint with attributes). This is to say that the maker of the image can signal, within a given cultural context, that the image is not to be looked at glancingly or casually".

In other words, this is the case when there is a special meditative or contemplative requirement, associated with some images. Throughout the discussion henceforth we will be referring to either one or another of these types, although the main emphasis will fall on the second type.

At this stage, I would only like to draw attention to the conceptual ambiguity of the statements, tending to deny the temporal element in pictorial arts, which makes me side with the view that the temporal-spatial distinction between the arts does not provide an adequate basis of categorization, or rather that it should be radically re-interpreted.

The Doctrine of the Purity of Art

One of the implications of the distinction between spatial and temporal arts comes down to the notion of the purity of the arts. As Rensselaer Lee claims, "no one will deny the general rightness of his [Lessing's] contention that the greatest painting like the greatest poetry, observes the limitations of its medium; or that it is dangerous for a spatial art like painting to attempt the progressive effects of a temporal art like poetry".[38] Clement Greenberg's essay 'Towards a Newer Laocoon' (1940) shares the same basic position. The author sees the history of

---

36 *op. cit.*

37 See Martin Kemp, *Spectacular Bodies: The Art and Science of the Human Body from Leonardo to Now*, (London, Berkeley, 2000).

38 Lee, '*Ut Pictura Poesis:* The Humanist Theory of Painting', 21.

avant-garde painting as that of a "progressive surrender to the resistance of its medium".[39] Modern painting "got rid of imitation, of 'literature'". In terms of pictorial perspective, the avant-garde breaks away from mathematically constructed space and affirms the flatness of the picture plane (we should remember here the alleged function of standard, linear perspective as fixing a moment in time). Its great achievement lies, according to Greenberg, exactly in that – keeping to the means of expression appropriate to its medium, as "it is by virtue of its medium that each art is unique and strictly itself".[40] In the nineteenth century painting had sunk low, in the author's estimation, as it had strived for an imitation of the effects of the dominant art form of the time – literature. "The damage", Greenberg says, lies not so much in "realistic imitation itself" as in "realistic illusion in the service of sentimental and declamatory literature".[41] What is relevant to our purposes is the author's general contention that there are certain effects appropriate for an art and therefore when attempting at effects not belonging by nature to this art's domain the result is invariably bad. We encounter the same argument, though a different context, with Sir Joshua Reynolds (1723–1792). In Discourse VIII, he says that "no art can be grafted with success on another" and most importantly, the painter "must depart (from poetry, as well as) from nature for a greater (plastic) advantage".[42] For the same reasons Walter Pater deplored coloured sculpture. "The use of colour in sculpture", Pater says in his essay on Luca della Robbia from 1872, "is but an unskilful contrivance to effect, by borrowing from another art, what nobler sculpture effects by strictly appropriate means".[43]

The doctrine of the purity of the arts has often existed alongside an analogy between painting and poetry that hierarchically evaluates them and favours one at the expense of the other.[44] Actually, literature has at times been assigned a higher place than painting based on the supposition that it can achieve some of the effects of the latter while the opposite would not be true. Associating himself with a long classical tradition, in *The Critic as Artist*, Oscar Wilde expresses this view in the following way: "while the poet can be pictorial or not, as he chooses, the painter must be pictorial always. For a painter is limited not as to what he sees in nature,

---

39 Clement Greenberg, 'Towards a Newer Laocoon' (1940) in his *The Collected Essays and Criticism*, (Chicago and London, 1986), vol. 1, p. 34.

40 Ibid., p. 32.

41 Ibid., p. 27.

42 Joshua Reynolds, *Discourses*, (London, 1905), p. 364.

43 Walter Pater, 'Luca della Robia' (1892) in his *The Renaissance*, (Oxford and New York, 1986, first in 1873), p. 42. Pater's position is more flexible, though, as in another piece, he says that "in its special mode of handling its given material, each art may be observed to pass into the conditions of some other art" and it is through "a partial alienation from its own limitations" that the arts "reciprocally [...] lend each other new forces" (Pater, W., 'The School of Giorgione' (1877) in his *The Renaissance*, p. 85).

44 See Lee, '*Ut Pictura Poesis*: The Humanistic Theory of Painting', 197–269.

but to what upon the canvas may be seen".[45] In a language, reminiscent of Hegel, Wilde says: "art is mind expressing itself under the conditions of matter".[46]

In the face of the acknowledged pre-eminence of poetry, Leonardo turned around the argument to prove the priority of painting. It is the poet who "may wish to rival the painter",[47] but fails to do so, because, as we saw, the "voids" between words in poetry "(dismember) the proportions"[48] and thus disrupt harmonious beauty. It is interesting to notice that Leonardo's position remains within the context of the antique division of the liberal (grammar, dialectic, rhetoric, arithmetic, geometry, astronomy and music) and mechanical arts. It is true, Leonardo says, that "poetry arises in the mind and the imagination of the poet",[49] "but imagination cannot see with such excellence as the eye".[50] While later the romantics would exalt the imagination, Leonardo's defence was built along entirely different lines, centring on emphasizing the role of the sense of sight above that of hearing.[51]

The immediate value of such hierarchical comparisons between poetry and painting may be doubtful to a modern reader. One indirect merit, however, they did have. It seems to me that the first step towards a systematic questioning of the time-space distinction between the arts was taken with the theme that one art could imitate the effects of another (and so was seen to have an advantage over the other).

The Problem of Pictorial Time in the Icon: Florensky and Remembering "Things that Happened the Week after Next"

Even though most twentieth-century scholars would hardly accept Lessing's distinction between the arts without qualification, the problem of pictorial time has remained somewhat peripheral to visual studies. Indeed, whatever discussion there has been seems to be mostly confined to the case of mathematically constructed space, which had been seen since Alberti to exclude the moment of the time in a

---

45 Oscar Wilde, 'The Critic as Artist' in *Oscar Wilde's Plays, Prose Writings, and Poems*, (London and New York, 1966), p. 30.

46 Ibid., p. 41. Most probably, Wilde had Pater in mind in this passage and thus recalls Hegel but via Walter Pater.

47 Kemp. (ed.), *Leonardo on Painting*, p. 24.

48 Ibid.

49 Ibid., p. 22.

50 Ibid.

51 On the Renaissance attitude to imagination and related notions, see Martin Kemp, 'From *Mimesis* to *Fantasia*: Quattrocento Vocabulary of Creation, Inspiration and Genius in the Visual Arts', *Viator*, 8 (1976): 347–99. On the whole, according to Kemp, Renaissance texts on art reveal an adherence to the ideal of *mimesis*, rather than to notions of individual creativity (347).

decisive manner.[52] Yves Bonnefoy's article "Time and Timelessness in Quattrocento Painting" is exceptional in that it treats of two fundamentally different temporal conceptions, underlying perspective and non-perspective images respectively.[53] Some of the authors who broach the issue of pictorial time do so by drawing notions from Maurice Merleau-Ponty's phenomenology. The most recent example is Paul Crowther's excellent "Eternalizing the Moment: Artistic Projections of Time".[54]

On the background of relatively small systematic interest in the topic of pictorial time in Western scholarship, it is striking to notice the persistent interest in the subject of pictorial time among Russian artists and writers in the first quarter of the twentieth century. Several interrelating factors contributed to the Russian preoccupation with time in the visual arts. On the one hand, in Russia, as elsewhere in Europe, the notion of the fourth dimension, frequently interpreted as time, had gained popularity by 1915. Linda Henderson has discussed the profound importance of the fourth dimension for a number of Russian avant-garde artists, such as Matyushin, Malevich, El Lissitsky, etc.[55] On the other hand, the reception of Cubism in Russia must have attracted still further attention, so far as the Cubists had broken with three-dimensional construction of space and their experiments with space were interpreted in the context of the fourth dimension. Guillaume Apollinaire's 1912 article on Cubism in *Les Soirées de Paris* was well-known, while in 1913 there were two Russian translations of Gleizes' and Metzinger's *Du Cubisme* (1912). Both texts make the connection between Cubism and the fourth dimension. Finally, Einstein's General Relativity Theory and Minkowski's "space-time" must have had an impact, too. Having in mind the scientific background of some of the Russian authors writing on the arts, Einstein and Minkowski might have been accessible even earlier than in the West. If we live in a space-time continuum, all objects we encounter, artistic or otherwise, exist in a unity of spatial and temporal characteristics. Thus, it is no longer possible to think of an artwork

---

52 Lew Andrews' discussion of continuous narrative (i.e., the representation of the same figure more than once in a painting) in the Renaissance deserves mention (Andrews, *Story and Space in Renaissance Italy: The Rebirth of Continuous Narrative*, as well as John Shearman's notion of the "transitive mode" (John Shearman, *Only Connect ... Art and the Spectator in the Italian Renaissance*, (Princeton, 1992). Leonardo's *Last Supper*, the common model of rigorous linear perspective, has been reconsidered by Leo Steinberg (Leo Steinberg, 'Leonardo's *Last Supper*', *The Art Quarterly*, 36 (1973): 392–410) and Martin Kemp (Kemp, M., *Leonardo da Vinci: The Marvellous Works of Nature and Man*, (London, Melbourne, 1981), p. 190).

53 Yves Bonnefoy, 'Time and Timelessness in Quattrocento Painting', in Norman Bryson, (ed.), *Calligram: Essays in New Art History from France*, (Cambridge, New York, 1988).

54 Paul Crowther, 'Eternalizing the Moment: Artistic Projections of Time', in his *Defining Art, Creating the Canon*, (Oxford, 2007), pp. 205–35.

55 Linda Henderson, *The Fourth Dimension and Non-Euclidean Geometry in Modern Art*, (Princeton, 1983), especially Chapter 5.

exclusively in terms of either spatial or temporal features. This was the general import of developments in physics.

These intellectual developments will be discussed in more detail later on, but at this stage it is interesting to see that they came together and were further elaborated in work done at RAKhN (The Russian Academy for Artistic Sciences) in the early 1920s. At the Psycho-Physical Department, founded by Vasilii Kandinsky, the topics of time and space in the visual arts were studied in conjunction from various perspectives. A list of papers delivered in the period 1921–1924 reveals that Malevich spoke on colour, light, and pointillism in time and space, while Natalia Kovalenskaia gave a talk on time and space in the fine arts.[56] At the Philosophy Department, which boasted among its members the religious philosophers Nikolay Berdyaev and Semeon Frank, the problem of time and space in the various arts was a mainstream topic. Boris Vipper, another member of RAKhN, would later go back to issues discussed at the Academy in his article 'Problema vremeni v izobrazitel'nom iskusstve' (The Problem of Time in the Visual Arts) (1962). Vipper explicitly started with the question if it was possible at all to speak of a category of time in the plastic arts.[57] The author's reply is that images in the plastic arts "have the ability to express absolutely concrete ideas of time".[58]

This is the broad intellectual background of Pavel Florensky's own theory of time in the icon. Florensky's notion of "reverse time" (*vremia obrashchennoe*, literally "time, which has been reversed")[59] is the only serious attempt so far, that I am aware of, of proposing a model of pictorial time in the icon. It was exposed mainly at the beginning of the *Iconostasis* (prepared in 1922)[60] and mentioned on several occasions in 'Mnimosti v geometrii' (Fictions in Geometry) (1922), at the time when the author was working at RAKhN. The main ideas in the *Iconostasis* were, however, the result of earlier work, as they were outlined in Florensky's lectures at the Theological Academy in Moscow in 1918.

According to Florensky's commanding thesis, reverse time is the temporal conception underlying dreams and art. In my discussion I will be using the latest Russian edition of Florensky's text of the *Iconostasis* in *Khristianstvo i kul'tura*

---

56 Boris Mikhailov, 'Coming of Age: Russian Art History as a Professional Discipline in the 1910s and 1920s', *Experiment*, 3 (1997): 10.

57 Boris Vipper, 'Problema vremeni v izobrazitel'nom iskusstve' (The Problem of Time in the Visual Arts) in *50 let Gosudarstvennomu muzeiu izobrazitel'nykh iskusstv imeni A.S.Pushkina. Sbornik statei* (A 50 Year Anniversary of the State Museum of the Visual Arts A.S.Pushkin. A Collection of Articles), (Moscow, 1962), pp. 134–50; rpt. *Statii ob iskusstve* (Articles on Art), (Moscow, 1970), p. 313.

58 Ibid.., p. 314; the translation is mine.

59 Pavel Florensky, 'Ikonostas' (Iconostasis) in *Khristianstvo i kul'tura* (Christianity and Culture), (ed.) A. Filonenko, (Moscow, 2001), p. 522; the translation is mine; this phrase is absent from the English translation.

60 There is an abbreviated version in English, entitled 'Icon', *The Eastern Churches Review*, 8 (1970) and the full translation from 1996, which I will be referring to.

(Christianity and Culture) (2000). The English translation (1996) is not always precise, at times even incorrect, so I will use my own translation, when I think it necessary.

I will be suggesting that Florensky's "reverse time" betrays a conceptual ambiguity, partly due to the terminology which, most probably, was chosen to fit the theory of "reverse perspective", of which Florensky was the most influential exponent and which will be discussed in the following chapter.[61] In other words, if we accept "reverse perspective" as the spatial co-ordinate of the pictorial world of the icon, "reverse time" will provide the temporal one. The problem with the term is that it suggests a literal reversal in the *direction* of time. At the same time, what is valuable in Florensky's discussion, I think, are those passages, dealing with the change in the *speed* of time which is fundamentally a reversal in kind. This latter aspect brings Florensky's argument close to my own, outlined in Chapter 4. It is also close in spirit to a notion, mentioned but, unfortunately, not pursued in a short text by Konstantinos Kalokyris.

In his article 'Byzantine Iconography and 'Liturgical Time', Kalokyris makes use of the concept of "liturgical time" in the sense of "condensed" or "concentré" time, as it is understood by some modern scholars.[62] The author suggests that Byzantine iconography reflects exactly this notion of "liturgical time".[63] However, he proves unable to support his thesis with convincing visual analysis and offers little beyond the vague statement that "sacred persons and events are made contemporary",[64] based on the assumption that the "overcoming of time" has the result of making "an event contemporary with us".[65] At the same time, it is not at all clear how Byzantine images actually incorporate this timeless notion. The article by Kalorykis is valuable, however, in that it poses a possible starting ground for the discussion. It is one of the few attempts so far to point explicitly to a connection between iconic art and the notion of timelessness, even though it fails to show the nature and structure of this connection.

Pavel Florensky's interest in the problem of time in the icon is much more persistent. Pictorial time is, for Florensky, an aspect of the larger question of the temporal dimension of the Eastern Orthodox world-view. The theme of time is recurrent throughout Florensky's writings in general,[66] which shows the author's preoccupation with it and the importance he attaches to it.

---

61 See Chapter 3.

62 Konstantinos Kalokyris, 'Byzantine Iconography and "Liturgical Time"', *Eastern Churches Review*, 1 (1966–1967): 359–63. On this interpretation of liturgical time, see Kalorykis's bibliography and also the section on timeless eternity in Chapter IV in this book.

63 Ibid., 360.

64 Ibid.

65 Ibid., 361.

66 See, for example, Florensky, *The Pillar and Ground of Truth*, in the context of the his overall religious world-view.

The term Florensky chooses to use suggests a procedure very much the same as the one involved in the working out of the concept of "reverse perspective", which supposedly reverses the laws of linear perspective. No one, to my knowledge, has pointed out the connection between the two notions, "reverse time" and "reverse perspective", which I believe to be important. On the one hand, both terms point to the idea of reversal, something that in both cases is due to and leads to misconceptions. On the other hand, there is an underlying implication of spatial-temporal unity and the idea that a certain conception of time leads to a certain conception of space and vice versa. As we saw, once this notion is applied to pictorial art, we start asking questions about the temporal dimension, incorporated in the structure of a work of art. For Florensky, though, the problem of pictorial time is important as it relates directly to his overall world-view.

Being in Russian religious philosophy at large is understood in Platonic and Neoplatonic terms. This has been mentioned on numerous occasions.[67] True reality resides in the other-worldly, while in this world we try to catch glimpses of it. In a short essay, Florensky discusses the term "realism" exactly in such Platonic terms and as a result, warns against the confusion in terminology – between "realism" and "naturalism" and "realism" and "illusionism".[68] Realism in general is defined as "a kind of tendency that affirms some kind of *realia* or realities – in contrast to illusions".[69] Artistic realism in particular is about "[uniting] us with the realities that are inaccessible to our senses".[70]

To perform this act of unity with the true being, a great transformation happens at the border-line of the two worlds – our existential categories are transformed into the categories of a higher being[71]. Logically speaking, a transformation can be realized in at least two ways – a qualitative change in kind, as, for instance from time to a timeless eternity or one that reverses the order of the phenomenon on the existing terms. The term "reverse time" implies that the latter is what happens in Florensky's conception of time – linear, Kantian time becomes reverse time. Reverse time seems to remain linear but its arrow points in the opposite direction. I will attempt a closer reading of Florensky's essay to see if this impression is true.

---

[67] In the case of Florensky's adherence to the Platonic tradition see Viktor Bychkov and Ilosha Kish, 'Vidna li neveroiushtemu Troitza Rubleva?: K voprosu o poniiatii ikonnogo znaka v '*Ikonostase*' P. Florenskogo' (Is Rublev's Trinity Visible to the Non-believer?: On the Question of the Iconic Sign in P. Florensky's *Iconostasis*) in Michael Hagemeister and Nina Kauchtschischwili, (eds), *P. A. Florensky i kul'tura ego vremeni* (P. A. Florensky and the Culture of His Time), (Marburg, 1995), pp. 399–411.

[68] Pavel Florensky, 'On Realism' in his *Beyond Vision: Essays on the Perception of Art*, ed. N. Misler, (London, 2002), p. 80.

[69] Ibid.

[70] Ibid., p. 181.

[71] I have discussed the boundary between the two worlds in Florensky's writings in a section of Chapter 4.

Art is mentioned by Florensky in that context as a "border-line case", alongside with dreams. Both dreams and art belong to the realm between the two worlds. That is, not all dreams but only mystical, morning dreams and not all art, but only "true" art, the supreme example of which is the ancient Russian icon.

To see art as the meeting of the two worlds was a common idea among Russian thinkers at the time. In an essay, entitled 'The Two Worlds in Ancient Russian Icon-painting' (1916) Evgeny Trubetskoy defines the icon as the "real contact between the two worlds of being, the levels of being – the heavenly one of calm and the earthly one, sinful, chaotic but striving towards repose in God existence".[72] Just as common was the interest in dreams and psychology stirred by the writings of Freud (*The Interpretation of Dreams* was published in 1900). Before psychology fell under the category of a "bourgeois science" in the late 1920s, it enjoyed a huge boom in the early years of the decade, as Alexander Etkind shows in his book *Eros nevozmoznogo: Istoriia psikhoanaliza v Rossii* (The Eros of the Impossible: History of Psychoanalysis in Russia).[73] Nicoletta Misler also mentions the fascination with Freudian psychology at the beginning of the century in Russia and in particular among intellectuals at RAKhN.[74] Alexei Siderov (1891–1978), a colleague and friend of Florensky's at the Academy, had spent time in Germany, where he had met Kandinsky. Siderov and Kandinsky would attend debates on Freud at the Café Stefan in Munich. At the Academy, Siderov delivered a paper in 1924, entitled "The Artistic Creativity of Dreams", which expressed these interests.[75]

The connection between dreams and art was, of course, an age-old theme, going back to Plato's *Timeaus*. It was also made by Freud, who maintained that "the psychical material out of which dreams are made"[76] is comparable to the material of visual art. Florensky himself seems to be relying heavily on pre-Freudian authors.

---

72 Evgeny Trubetskoy, *Umozrenie v kraskakh: tri ocherka o russkoi ikone* (A Contemplation in Colour: Three Essays on the Russian Icon), (Paris, 1965, first in 1915–1918), p. 63, the translation is mine. There is an English translation in Evgeny Trubetskoi, *Icons: Theology in Colour*, (New York, 1973).

73 Alexander Etkind, *Eros nevozmoznogo: Istoriia psikhoanaliza v Rossii* (The Eros of the Impossible: History of Psychoanalysis in Russia), (Moscow, 1994). There is an English translation from 1997.

74 Nicoletta Misler, 'Toward an Exact Aesthetics: P. Florensky and the Russian Academy of Artistic Sciences', in John Bowlt and Olga Matich, (eds), *Laboratory of Dreams: The Russian Avant-garde and Cultural Experiment*, (Stanford, CA, 1996), p. 125. Interestingly, some members of the Bolshevik elite were attracted to psychoanalysis for a while, too. The People's Commissariat of Enlightenment had a psychoanalysis department, which sponsored a "children's home laboratory" in Moscow, attended, among others, by Stalin's son.

75 Alexei Siderov, 'The Artistic Creativity of Dreams', *Experiment*, 3 (1997): 179–80.

76 Sigmund Freud, *The Interpretation of Dreams* (New York, 1965), p. 347.

He explicitly mentions F. W. Hildebrandt,[77] who was writing in Leipzig in the late nineteenth century. Actually, according to the contemporary Russian scholar Alexander Mikhailov, the closest in spirit to Florensky and a possible influence on him is G. H. Schubert (1780–1860), a natural philosopher from the first half of the nineteenth century.[78] In his works, *Ansichten von der Naturwissenschaft* (Views on Natural Science) (1808) and *Des Symbolik des Traumes* (The Symbolism of Dreams) (1814), Schubert puts forward his thesis that the ability to reach the invisible and the world beyond is the natural, physical and spiritual condition of man as an element in the hierarchy of being. Schubert claims that this is exactly what happens in dreams. It is worthwhile noting the author's contention that the boundary between the two worlds is temporal. In dreams, consequently, taking into consideration their function in Schubert's theory, there is a time flow, different from the one we are aware of in woken consciousness.

What Florensky attempts to do is to work out an idea of time, within that overall intellectual framework, that describes the time of dreams and art. At the point where the two worlds meet and where the transformation of categories occurs reside dreams and art which give us an idea of true being.

The notion of reverse time is illustrated by Florensky through the composition of the dream. It should be noticed that the author's preoccupation with dreams is different from Freud's. Freud speaks of two types of dreams – the ones due to external stimuli and the ones due to internal stimuli and his interest is mainly excited by the latter. Florensky, on the contrary, concentrates on the former and actually states at one point that all dreams are due to external stimuli.[79] That dreams are the result of external stimulation is a typical nineteenth-century idea which lies at the heart of the physiological theory of dreams.

Let us take a few instances of dreams due to external stimuli, cited by Freud, and analyze their composition in Florensky's terms. Freud actually derives his examples from L. F. Alfred Maury's *Le Sommeil et les Rêves* (1861), one of the proponents of the physiological theory:

> "He was given some eau-de-cologne to smell. – He was in Cairo in Johann Maria Farina's shop. Some absurd adventures followed, which he could not reproduce. He was pinched lightly on the neck. – He dreamt he was being given mustard plaster and thought of the doctor who had treated him as a child [...] A drop of

---

77 The name is misspelled in the English translation (see Pavel Florensky, *Iconostasis*, p. 38).

78 Alexander Mikhailov, 'O. Pavel Florensky kak filosof granitsy' (Father Pavel Florensky as a Philosopher of the Boundary), *Voprosy iskusstvoznaniia* (Problems in the Theory of Art), 4 (1994): 42.

79 Florensky, 'Ikonostas' (Iconostasis) in *Khristianstvo i kul'tura* (Christianity and Culture), ed. A.Filonenko, (Moscow, 2001), p. 523; Pavel Florensky, *Iconostasis*, (Crestwood, New York, 1996), p. 35.

> water was dropped on his forehead. He dreamt he was in Italy, was sweating violently and was drinking white Orvieto wine."[80]

The direct cause for the dream is an external stimulus A – an eau-de-cologne, a pinch on the neck, a drop of water. In the dream, however, A is interpreted as the symbolic image X – the shop in Cairo, the application of mustard plaster, sweating, and drinking wine. Unlike A, X belongs to the system of the dream and is the outcome of an event or a chain of events Y1, Y2, Y3, etc., happening in the dream. While A is the external cause, X is the spiritual cause for these events. The sequence of events in the dream has a certain direction and this is from effect to cause and from what follows to what precedes. Within the system of the dream X comes first and from it follow Y3, then Y2 and finally Y1. In other words, time is reversed in respect to everyday consciousness. The typical illustration is the ringing of the alarm-clock (A), which in the dream could be a shot (X). Our memory sorts out the information of a dream in a coherent story, the outcome of which is X. Within the dream, X precedes all the events that have caused it, while the events themselves unfold backwards from the future to the past. The so described composition of the dream in which the end determines the beginning and all other events is teleological:

> "its events occur because of its *denouement*, in such a way that the *denouement* will not be left hanging in the air but will, instead, exhibit deep programmatic rationality and is not an unhappy chance but possesses a deep practical motivation".[81]

Florensky sees the time flow in the dream as the exact opposite to that of Kantian time[82]. In the dream

> "[time] runs and runs acceleratedly towards the actual and against the movement of time in waking consciousness. Dream time is *turned inside out*, which means that all its concrete images are also *turned inside out* with it: and that means we have entered the domain of imaginary space".[83]

Dreams are the first and the simplest (in the sense that we are accustomed to them) stage of life in the invisible. Later, in memory they are arranged according to the temporal order of our world. In the 'Transcendental Aesthetic' in the First Critique Kant describes this temporal organization of the material, provided by the senses. Initially, the raw material of sensible experience arrives in human consciousness without structure. The mind then proceeds to order it by imposing

---

80 Freud, *The Interpretation of Dreams*, p. 25.

81 Florensky, 'Ikonostas' (Iconostasis), p. 525; Florensky, P. , *Iconostasis*, p. 38.

82 Ibid., p. 527; Ibid., p. 41.

83 Ibid., p. 528; Ibid., p. 41.

a temporal and spatial framework over it. The organization of experience follows a process of synthesis – a synthesis of apprehension in intuition and a synthesis of recognition in a concept, as human knowledge requires both having intuitions and applying concepts.

So far it appears that "reverse time" is indeed reverse in terms of the direction in which it unfolds, which is opposite to the one we are aware of in everyday life. Another characteristic of Florensky's conception of time in dreams and art challenges this impression. Reverse time, Florensky claims, has its own *duration*, as well. What from an outside, empirical point of view may seem relatively short in time (minutes, hours) inside dreams might cover a much longer period (years, decades). As Thomas De Quincey (1785–1859) lucidly describes his own dreams in *The Confessions of an English Opium Eater*: "sometimes I seemed to have lived for seventy or a hundred years in one night".[84] De Quincey's experience might partly be due to the effect of opium, but his "powerfully affected"[85] senses of space and time could serve as indications to what generally happens in dreams. Actually, Florensky says, the speed of time in dreams might be infinite and reverse in respect to empirical time: "few have considered the possibility of an infinite speed of time flow and even, turning inside out in respect to itself, in the transition to infinite speed to gain a meaning reverse (*obratnii*) in its flow".[86]

It is that moment of reverse time which is of interest for the discussion here. Florensky describes it as the teleological aspect of time in the invisible:

> "time, indeed, can be an instant, from the future to the past, from the effects to the causes; time can be teleological, especially when our life makes the transition from the visible to the invisible, from the actual to the illusionary".[87]

It becomes obvious at this point that there are two interrelated features of reverse time – one is its direction, the other its duration. The latter can be "infinite in speed" and "an instant", both of which come down to the same thing. Thus, the phenomenon we are talking about is *different in kind*, not just reverse to the time in the woken consciousness. For this reason, I suggest, the term "reverse time" could be very misleading. Florensky himself often provides grounds for such misconceptions and for a simplistic and, I think, incorrect understanding of reverse time as, for instance, when he describes the invisible world as "God's

---

84 Thomas De Quincey, *Confessions of an Opium Eater*, (Harmondsworth, 1971, first in 1821), p. 104.

85 Ibid., p. 103.

86 Florensky, P., 'Ikonostas', p. 522; the translation is mine; the English translation for "*obratnii smisl' svoego techeniia*" is "time that flows backwards" (p. 35) is incorrect, as it implies only a change of direction, while Florenskay explicitly says "smisl'", i.e., "meaning", which could imply other things, besides direction.

87 Ibid.; my own translation: among other things, the English translation skips "*teleologicheskim*", i.e., teleological (p. 35).

creation, but (beheld) from the *other* side".[88] Particularly unfortunate is the example with a friend's dream. In his dream, Florensky's friend gets to know that under the surface of the earth there is grass and trees growing, but they grow upside down, with their roots upwards and their leaves pointing downwards. We are reminded of Alice going down the rabbit-hole and thinking: "I wonder if I shall fall right *through* the earth! How funny it'll seem to come out among the people that walk with their heads downwards!".[89] With Florensky this graphic description is meant to illustrate the notion that the invisible world is reverse to ours. It is, however, misguided as the image it evokes could easily give rise to a concept of reversal in the narrow sense of the word, where all the elements of a phenomenon become their exact opposites. In other words, reversal could be understood as an inversion, i.e., as a turning upside down of a phenomenon. The element of duration surely leads to something more than that and namely, to the notion of an in-depth reversal of kind. The very reversal of time suggests actually a lack of duration – this is exactly what "infinite in speed" and "instant" imply. The lack of duration, on the other hand, is an aspect of the concept of timelessness (see Chapter 4). In this sense, Florensky's view is similar in spirit to the one Kalokyris was driving at. It makes one think of Maurice Maeterlinck's (1862–1949) description of our experience, when "we discover with uneasiness that time, on which we base our whole existence, itself no longer exists".[90] We come to feel that "all that comes and all that goes passes from end to end of our little life without moving by a hair's breadth around its motionless pivot".[91]

When Florensky claims a possible scientific significance of his theory of reverse time, he reverts the argument to a completely different direction, which is not of direct relevance to our interests here. The next stage after the discovery of instant time by Karl Duprel, and consistent with Duprel's notion, should be the discovery of reverse time, Florensky believes. It is, however, noticeable even to non-specialists that Florensky was building his notion of reverse time on the background of contemporary research, associated with Einstein. According to Einstein, Newton's laws of motion were no longer applicable when we deal with bodies moving at high speed (i.e., faster than the speed of light in empty space).[92] Einstein is not talking here of infinite speed, as even the speed of light is finite,

---

88 Ibid., pp. 528–9; Ibid., p. 42.

89 Lewis Carroll, 'Alice's Adventures in Wonderland' in Lewis Carroll, *The Penguin Complete Lewis Carroll*, (Harmondsworth, 1982), p. 17. Interestingly, Florensky discusses Lewis Carroll for his mathematical work (See Pavel Florensky, *The Pillar and Ground of the Truth*, tr. and annotation by B. Jakim, (Princeton, New Jersey, 1997), Chapter: 'A Problem of Lewis Carroll and the Question of Dogma', pp. 355–9.

90 Maurice Maeterlinck, 'The Knowledge of the Future' in his *The Unknown Guest*, (London, 1914), p. 124.

91 Ibid., pp. 124–5.

92 See John D. Barrow, *Impossibility: The Limits of Science and the Science of Limits*, (London, 1998), pp. 24–5.

but speed becomes a criterion for distinguishing between two systems, adhering to different laws. Florensky's distinction between time in dreams and art and time of the woken consciousness, too, is based on the factor of speed and not simply on a difference in direction, as his term might suggest at first glance. It is interesting to notice that in the year Florensky prepared his text the German scientist Hermann Weyl, wrote: "it is not impossible for a world-line (in particular, that of my body) although it has a time direction at every point, to return to the neighbourhood of a point, which it has already passed through".[93] What is at issue here is a conception of time that is cyclical and non-directional (other than locally) in distinction to linear and directional time. I believe that Florensky's notion of "reverse time" is analogous and therefore, the key to it is the distinction between directional and non-directional time, rather than between linear and reverse.

While Florensky's anticipated discovery of reverse time has not been realized scientists, philosophers of science and novelists continue to be fascinated with the possibility of time travel and are intrigued by the question: "Is a reverse world logically possible?"[94] In a sense, we still wonder with Alice at the White Queen, who remembers best "the things that happened the week after next".[95]

What is important for the purposes of this essay is not the scientific significance of Florensky's ideas on reverse time, but the very attempt by the Russian writer to discover a temporal dimension underlying the medieval icon. Florensky does so from the position of a presupposed interconnectedness between pictorial time and pictorial space. His idea of the spatial construction of the icon is largely responsible, in my opinion, for his terminology and occasional passages which tend to see categories in terms of the "reversal" of existing ones. The main idea, as I see it, is to propose a conception of the time at the borderline between the two worlds which realizes the transformation from the profane to the sacred, from the visible to the invisible. I have tried to bring across my conviction that reversibility in this case should be understood as genuine, qualitative *difference* and not just a reversal in the direction of linear time.

The Problem of Time in the Pictorial Arts – Bakhtin's "Chronotope"

A bit more than a decade after Florensky's writings on "reverse time", Mikhail Bakhtin (1895–1975) published his essay 'Forms of Time and of the Chronotope in

[93] Hermann Weyl, *Space, Time, and Matter*, (London, 1922). Strictly speaking, Weyl means a cyclical return, not literally a reversal, which would imply a linear one.

[94] Roger Teichmann, *The Concept of Time*, (Basingstoke, 1995), p. 168. The same and related issues are the subject of Hans Reichenbach's book *The Direction of Time*, (Berkeley, 1956) and of a chapter of the same title in William Newton-Smith's of *The Structure of Time*, (London, 1980).

[95] Lewis Carroll, 'Through the Looking Glass', p. 181.

the Novel' (1937–38),[96] which introduces the term "chronotope". Bakhtin belonged to a rather different intellectual set than Florensky. The latter moved among the crème de la crème of the intelligentsia, while Bakhtin – who was to become probably Russia's best-known thinker in the West – was much more isolated at the time. However, Bakhtin, too, was exposed to the prevailing ideological currents in early twentieth century Russia and the "chronotope" was another way of stating that the work of art is a unified spatial-temporal construct, just as any other object of experience. The reason Bakhtin is of interest in the context of our discussion is that, I believe, his influential concept grew largely out of Russian concerns in the first quarter of the twentieth century, which we mentioned in the case of Florensky. Two ideas, however, become even more explicit with Bakhtin – the Einstein connection and the Kant connection. Einstein's authority was evoked to insist on the inseparability of the spatio-temporal construct. Kant was referred to in order to stress the significance of time and space as coordinates of a world (here, of an artistic world).

The Russian author defines the "chronotope" in the following way: "We will give the name *chronotope* (literally "time space") to the intrinsic connectedness of temporal and spatial relationships that are artistically expressed in literature".[97] Bakhtin acknowledges that the term derives from Einstein, but he says that "the special meaning it has in relativity theory is not important for our purposes", as it is borrowed by literary criticism "almost as a metaphor (almost but not entirely)".[98] He emphasizes that the main importance of Einstein and his use of space-time terminology is, for aesthetics, the notion of "the inseparability of space and time (time as the fourth dimension of space)".[99] While Bakhtin employs the chronotope exclusively as a category of literary criticism and states that he will not be applying it to other spheres of culture, his descriptions of it often carry more general overtones as in:

> "In the literary artistic chronotope, spatial and temporal indicators are fused into one carefully thought-out, concrete whole. Time, as it were, thickens, takes on flesh, becomes artistically visible; likewise, space becomes charged and responsive to movements of time, plot and history."[100]

---

[96] Mikhail Bakhtin, 'Forms of Time and of the Chronotope in the Novel' in Mikhail Bakhtin, *The Dialogical Imagination: Four Essays by M.M.Bakhtin*, (ed.) Michael Holquist, (Austin, TX, 1981).

[97] Ibid., p. 84.

[98] Bakhtin has often been attacked for his, sometimes, imprecise use of terminology. Describing his own term as "almost but not entirely (a metaphor)" is surely quite confusing.

[99] Bakhtin, 'Forms of Time and of the Chronotope in the Novel', p. 84.

[100] Ibid.

Put in those terms, the category of the chronotope could be stretched out to cover a larger field and it would make sense to apply it to painting as well. It is indeed surprising that there have been no serious attempts to do so. Deborah Haynes's *Bakhtin and the Visual Arts* stands largely on its own in the attempt to assess the relevance of Bakhtin's ideas in the field of painting and sculpture. Haynes, however, concentrates mainly on Bakhtin's theory of creativity and related issues and only passingly mentions the concept of the chronotope.[101] There is also an article by Jay Ladin which considers the usefulness of the chronotope in the analyzes of the other arts outside literature. The author believes that "the formal language of the chronotope could be significantly enriched by surveying the existing criticism of other media for insights regarding the construction of space and time".[102] Ladin pays particular attention to the relevance of the chronotope as a critical tool in the sphere of film, while Laurin Porter's article 'Bakhtin's Chronotope: Time and Space in *A Touch of the Poet* and *More Stately Mansion*'[103] is a concrete instance of its application in theatre studies. Ladin's thesis is – and this is important for our purposes – that as the relations between chronotopes are graphically demonstrable they would yield much easier to identification in non-verbal media. Thus, while the application of the chronotope to pictorial art might be a challenging task it could be very rewarding at least on two grounds. Firstly, the chronotope as "a powerful but underdeveloped critical tool"[104] might be better explicated and further developed by such an application, addressing its natural propensity for visualization. Secondly, the chronotope might help to illustrate a fundamental aspect of pictorial art which is the topic of the present chapter.[105]

Much critical attention, on the other hand, has been devoted to a footnote in 'Forms of Time and the Chronotope in the Novel', which, for this reason, deserves a full quote:

> "In his 'Transcendental Aesthetics' [one of the main sections of the *Critique of Pure Reason*] Kant defines space and time as indispensable forms of any cognition, beginning with elementary perceptions and representations. Here we employ the Kantian evaluation of the importance of these forms in the cognitive

---

101 See Deborah Haynes, *Bakhtin and the Visual Arts*, (Cambridge, 1995).

102 Jay Ladin, 'Fleshing out the Chronotope' in Caryl Emerson, (ed.), *Critical Essays on Mikhail Bakhtin*, (New York, 1999), p. 228.

103 Laurin Porter, 'Bakhtin's Chronotope: Time and Space in 'A Touch of the Poet' and 'More Stately Mansions', *Modern Drama*, 43/3. (1991): 369–82.

104 Ladin, 'Fleshing out the Chronotope', p. 230.

105 Important recent works on Bakhtin, his place in intellectual history and in particular the chronotope, as well as the Kantian and Neo-Kantian connections of the Russian philosopher are: Ken Hirschkop, *Mikhail Bakhtin: An Aesthetics for Democracy* (Oxford, 1999), Galin Tihanov, *The Master and the Slave: Lukács, Bakhtin, and the Ideas of Their Time* (Oxford, 2000), Craig Brandist, *The Bakhtin Circle: Philosophy, Culture and Politics*, (London and Sterling; Virginia, 2002).

> process, but differ from Kant in taking them not as 'transcendental' but as forms of the most immediate reality. We shall attempt to show the role these forms play in the process of concrete artistic cognition (artistic visualization) under conditions obtaining in the genre of the novel."[106]

Already in his main text, when Bakhtin defines the chronotope as a "formally constitutive *category* of literature" (the emphasis is mine) he implies a point of departure from Kant. As it is known, Kant did not include time and space in his table of categories. For Kant, just as for Aristotle, a category is something coming about as a result of thought rather than perception. On the other hand, time and space, Kant tells us in the 'Transcendental Aesthetics', are *a priori* forms of perception, i.e., they precede, are prior to experience.

It should be noted, that for all his differences from Kant, Bakhtin's approach owes a lot to the German philosopher. As Bernhard Scholtz says in his article "Bakhtin's Concept of 'Chronotope: The Kantian Connection'": "Bakhtin was considering the possibility of talking about varying forms of experiencing an object *within the context of* the conceptual framework of *(Neo-) Kantian* transcendental analysis, rather than rejecting transcendental analysis".[107] For the practical purposes of literary – and I should add, art historical – criticism, Scholz describes the main difference between Kant and Bakhtin in their field of concern. While Bakhtin analyzes chronotopes "reflected in art", Kant, it could be said, concentrated on "actual" chronotopes, i.e., prior to their artistic assimilation.[108]

From a strictly philosophical perspective, Bakhtin gives grounds for the belief that he confuses the empirical and transcendental analyzes.[109] It should be born in mind that he was dealing with one of the notoriously difficult passages by Kant, where Kant himself left ample grounds for misunderstanding. However, apart from this, Bakhtin was forcing an application of Kant to serve his own ends. According to Gary Morson and Caryl Emerson, the point he was making "is that

---

106 Bakhtin, p. 85. On this, see Michael Holquist and Katriona Clark, 'The Influence of Kant in the Early Works of M.M.Bakhtin' in Joseph Strelka, (ed.), *Literary Theory and Criticism: Festschrift for René Wellek*, (Berne and Frankfurt, *c.*1984), pp. 299–313; Gary Morson and Caryl Emerson, *Mikhail Bakhtin: The Creation of a Prosaic* (Stanford, 1990), pp. 367-9; Michael Holquist, 'Introduction: The Archetectonics of Answerability' in Michael Holquist and Vadim Liapunov, (eds), *Art and Aswerability: Early Philosophical Essays by M.M.Bakhtin*, (Austin, TX, 1990), pp. ix-xlix; Barnhard Scholz, 'Bakhtin's Concept of "Chronotope": The Kantian Connection' in David Shepherd, (ed.), *The Contexts of Bakhtin: Philosophy, Authorship, Aesthetics*, (Australia, Canada, 1998), pp. 141–73.

107 Scholz, 'Bakhtin's Concept of 'Chronotope', p. 149.

108 *op. cit.*, p. 151.

109 See Holquist and Clark, 'The Influence of Kant in the Early Works of M. M. Bakhtin'.

time and space vary in *qualities*: different social activities and representations of those activities presume different kinds of time and space".[110]

To what an extent the application of Kant's transcendental forms in an empirical context is philosophically defensible is a question that oversteps our immediate concerns. Bakhtin's own definition of philosophy is rather loose, as in the following "our analysis must be called philosophical mainly because of what it is not: it is not a linguistic, philological, literary or any other particular kind of analysis [...] (it moves) in spheres that are liminal, i.e., on the borders of the aforementioned disciplines".[111] What is important is that Bakhtin's concept of the "chronotope" is meant to describe the mutual interdependence of *artistic* time and space in the *work of art*. As Tzvetan Todorov mentions, Bakhtin's works are about the relation between text and world, where the text puts forward a model of the world, the constitutive elements of which are time and space.[112] Thus, each literary genre that Bakhtin analyzes, from the classical Greek novel to Rabelais, represents a specific spatio-temporal configuration, i.e., a specific chronotope. In this sense, in Bakhtin's genre theory, a genre becomes synonymous with chronotope. It could be said that the present book is partly about the working out of a chronotope of icon art, by attempting to illustrate the relationship between pictorial space, as represented by "reverse perspective" and pictorial time, underlying "reverse perspective".

The possibility of understanding the chronotope as a more general aesthetic category comes across in Bakhtin's analyzes of literary works. In literature not only time, but also space play a role in the overall structure of the work. For instance, in *Rabelais and His World*, the Russian critic discusses the chronotope particularly in relation to images in Rabelais. Ideas of time and space are closely fused in the interpretation of the grotesque human body. The body's spatial positionings build an image of the defeat of time and death.[113] Thus, when Joseph Frank, for instance, suggested that "spatial form" was a central category of literary modernism[114] he refers, even if implicitly, to a major theme of Bakhtin's writings. It seems to me that the field was left open to see the visual arts in terms of the chronotope and so explore their usually neglected temporal dimension.

---

110 Morson and Emerson, *Mikhail Bakhtin: Creation of a Prosaic*, p. 367.

111 Mikhail Bakhtin 'Problema teksta v lingvistike, filologii, i drugikh gumanitarnikh naukakh. Opit filosofskogo analiza' (The Problem of the Text in Linguistics, Philology and in the Other Humanities. An Attempt at Philosophical Analysis) in *Estetika slovesnogo tvorchestvo* (The Aesthetics of Linguistic Creativity), (Moscow, 1979).

112 Tsvetan Todorov, 'From *Mikhail Bakhtin: The Dialogical Principle*' in Caryl Emerson, (ed.), *Critical Essays on Mikhail Bakhtin*, (New York, 1999), p. 193.

113 Mikhail Bakhtin, *Rabelais and His World*, (Bloomington, IN, 1984).

114 Joseph Frank, 'Spatial Form in Narrative' in Jeffrey Smitten and Ann Daghistany (eds.), *Spatial Form in Narrative*, (Ithaca, 1981), p. 13.

Conclusion

In this chapter I have attempted to illuminate aspects of the problem of time in pictorial art. I have concentrated on two issues, worked out in reaction to the extreme interpretation of the *ut pictura poesis* tradition, which tends to almost identify the arts of painting and poetry. The other extreme position has stressed the differences between painting and poetry and I have been mainly interested in the spatial-temporal distinction. On the one hand, temporality has been denied to pictorial art on the basis that a picture can convey an impression in an instant, "in the twinkling of an eye". It is presumed that in this case, time is excluded and plays no role. On the other hand, it is assumed that the space-time distinction actually provides a proper framework that the arts should follow – each should not attempt the effects of any of the others.

While pictorial time is still an understudied problem in Western scholarship, it is rarely realized that it was a mainstream topic of study in Russia at the beginning of the twentieth century and especially in the 1920s. Two Russian authors were singled out – Florensky and his notion of the "reverse time" of the icon, and Bakhtin and his concept of the "chronotope", which may have important implications for the visual arts. In both cases, time in visual images can be seen as the direct result of the view of the work of art as existing in a spatial-temporal unity. This has meant that the interest in pictorial time went hand in hand with the interest in pictorial space.

Chapter 2
On Reverse Perspective – a Critical Reading

Within the overall course of world art over the ages, linear perspective has been the exception rather than the rule. Still, without even looking for a justification to do so, there still persists a tendency to describe other, fundamentally different systems of spatial representation in terms of linear perspective. Within the discourse of normative aesthetics, this approach leads to the familiar observation that non-linear perspectival representations are the result of "artistic inability". What is interesting, however, is that even opponents to normative aesthetics often adopt the same procedure, even though they do not reach, of course, the same conclusions. I will discuss in this chapter the specific case of the academic research, conducted so far on the problem of "reverse perspective".

I will be using the term "reverse perspective", in preference to "inverse", following the recent English translation of Florensky's essay "Reverse Perspective".[1] The term "reverse perspective" refers to the construction of pictorial space in icon art. One of the main theses I will put forward in my book is that "reverse perspective" is a *defining* feature of the icon. In other words, alongside the motley of more or less noticeable stylistic changes from Late Antiquity to the present, this principle of spatial, pictorial treatment has persisted to this day. Thus, the subject of our study is what Otto Pächt calls a "constant" or "something that remains unchanged amid the flux of phenomena and must therefore be highly substantial".[2]

In terms of what has been just said, the fact that the problem of "reverse perspective" has received very little academic attention acquires a special significance. We basically lack a theory for an important aspect of an art form which has been around for more than a thousand years, has deeply influenced the whole of European medieval art and has endured in the face of Renaissance perspective to become a symbol of Eastern Orthodox spirituality.

The relatively few studies, devoted to the problem of "reverse perspective" are mostly by Russian authors. I will be referring to the following works – Pavel Florensky's "Reverse Perspective" (read in 1920), Lev Zhegin's *Iazik zhivopisnogo proizvedeniia* (The Language of the Work of Art) (1970) and Boris Uspensky's *The Semiotics of the Russian Icon* (1971). I will attempt to show that Russian scholarship

---

[1] Gombrich, too, translates "Umkehr" as "reversal" (See Ernst Gombrich, *Aby Warburg: An Intellectual Biography*, (Oxford, 1986, first in 1970), p. 248, while Christopher Wood uses the term "reverse perceptive" in his translation of Panofsky's "Perspective as a Symbolic Form" (New York, 1997), p. 114.

[2] Otto Pächt, *Practice of Art History: Reflections on Method*, (London, 1999, first in German in 1986), p. 135.

on this issue proceeds from a definition of "reverse perspective", as worked out by Oscar Wulff in "Die umgekehrten Perspektive und die Niedersicht" (1907).[3] The only challenge to Wulff's position, of which I am aware, is Karl Doehlemann's alternative view in 'Zur Frage der sogenannten "umgekehrten Perspektive" (1910).[4] This brief bibliographical review comprises what I believe to be the most important writings on the subject.

This chapter will be concerned with the implications of the term "reverse perspective". Most importantly, however, the legitimacy of the term will be considered in view of what is "going on" in the icon on the level of artistic space. The appropriateness of the term will be critically appraised in view of the state of research so far and the need for a new approach to the problem will be considered.

Implications of the Term "Reverse Perspective": Reverse of What?

Before outlining briefly some of the major characteristics of reverse perspective it makes sense to consider the implications of the term itself. The first and most obvious implication of term "reverse" is that it has reversed something that already exists. It is thus suggested that this kind of pictorial representation reverses the rules observed by linear perspective. In that sense, the term is extremely misleading both on historical and philosophical grounds. Although there were perspectival systems of representation in Classical Antiquity, Eastern Orthodox art can hardly be thought as an immediate, deliberate and polemic response to forms of art so culturally remote. On the other hand, if "reversal" is defined with respect to Renaissance linear perspective, as generally happens, it is necessary to point out that "reverse perspective" had been in existence long before what it is supposedly reversing. Artists working in the Byzantine style, like all medieval painters before the fifteenth century,[5] knew nothing of strict, mathematical pictorial space. However, almost all work on reverse perspective has been done from that premise – reverse perspective is taken as systematically negating the basic assumptions of linear perspective. The overall approach proceeds from a definition in negative terms, which is always unsatisfactory to some degree.

An even more fundamental objection concerns the terms of conceptual reference. It is a basic tenet of philosophy that only phenomena belonging to the same category can be compared meaningfully. The decisive problem for the discussion here will be to determine whether the pictorial space of the icon can be adequately grasped

[3] Oscar Wulff, 'Die umgekehrten Perspektive und die Niedersicht. Eine Raumanschaungsform der altbyzantischen Kunst und ihre Fortbildung in der Renaissance' in *Kunstwissenschaftliche Beiträge A.Schmarsow gewidmet*, (Leipzig, 1907).

[4] Karl Doehlemann, 'Zur Frage der sog. "umgekehrten Perspektive"', *Repertorium fur Kunstwissenschaft*, (Berlin, 1910).

[5] Martin Kemp fixes the time of Brunelleschi's construction somewhere before 1413 (Marin Kemp, *The Science of Art*, (London and New Haven, 1990), p. 9).

as reverse perspective and if it could be taken as belonging to the same category as linear perspective. Is it "perspective" at all? In practice, the question would come down to whether the major characteristics of linear perspective can be legitimately opposed to those of iconic space and vice versa or whether we are dealing with two radically different pictorial phenomena, requiring different frames of optical and conceptual reference.

Studies on "reverse perspective" almost invariably proceed from an introductory account of linear perspective to an analysis of its "reverse" variant. The notions worked out in the process of analysis of Renaissance works of art are automatically deployed to explain icon art. This procedure is well exemplified in the line of Russian thought which I will be concerned with, starting from Florensky (first stage), via Zhegin (second stage) and reaching a logical conclusion with Uspensky (third stage).

If the pictorial space of icons and the space of Renaissance paintings could be proved to belong to the same category of visual functionality this procedure would be wholly justified. If not, the whole approach would ultimately lead down a blind alley. In that case, a new terminology and fundamentally new concepts should be worked out. The ultimate aim, signalled here, is to produce a convincing theory of the pictorial mechanisms of the icon on their own terms.

The First Stage: Florensky

Florensky wrote his essay "Reverse perspective" in 1919 in the form of a lecture which he delivered to the Byzantine Section of the Moscow Institute of Historical and Artistic Researches and Museology in the following year. In the next several years the topic of pictorial space persisted in attracting Florensky's attention. In his resume at RAKhN (The Russian Academy for Artistic Sciences) Florensky described himself as a "Professor in the Analysis of Space" and between 1921 and 1924 he taught courses on "The Analysis of Spacial Forms", "The Analysis of Perspective", etc. These interests, as we saw, fit well in the intellectual ambiance at RAKhN.

The fate of Florensky's text on "reverse perspective" was extraordinary and even dramatic. The essay was prepared for publication, but, for some reason, did not reach the press at the time. It came out for the first time in Russian only in 1967,[6] after it had been almost completely lost to the world for fifty years, as its author was considered politically suspect. Once it came out, however, it became the one most influential writing on the subject. Henceforth, especially in the Russian-speaking world, the term "reverse perspective" carried immediate associations with Florensky's name and provided the starting ground for any discussion on the problem.

---

[6] See Pavel Florensky, 'Obratnaia perspektiva' (Reverse Perspective), *Trudy po znakovim sistemam*, 3 (1967): 381–416.

In my exposition I will comment on certain aspects of Florensky's position which have proved to have an enduring impact on subsequent research. Throughout, I have tried to keep an eye on the way Florensky fits into the wider context of ideas, current at his time, as well as more recent studies. I will be referring to the latest Russian edition of the 'Reverse Perspective' in *Khristianstvo i kul'tura* (Christianity and Culture) (2001), while the English translation will be from the collection of Florensky's works on art, presented for the first time to the English-speaking public in *Beyond Vision: Essays on the Perception of Art* (2002). I will be concerned in particular with the second part of Florensky's work, dealing with the theoretical premises of perspective, and especially with sections XIV and XV.

Florensky outlines six pre-conditions for linear perspective which he holds to be false, an opinion justified by scientific research in the nineteenth century, especially in Germany and Austria. Reverse perspective does not subscribe to these preconditions and is thus defined in terms of their negation.

The first and basic premise, from which all others proceed, is the one of Euclidean space. Linear perspective is based on the notion that we live in a Euclidean three-dimensional space, of which our vision gives us instances and to which instances pictorial perspective gives a permanent visual form. However, Florensky says, three distinct levels have to be distinguished in the problem of space, of which abstract, geometrical space is only a particular case. There is a level of physical space and there is no reason to say that it is Euclidean.[7] When discussing the level of physiological space Florensky uses Ernst Mach (1838–1916), a professor of psychology at Vienna University and one of the foremost philosophers and physicists at his time, as a main source. Physiological space is not Euclidean either and, according to Mach: "If we accept that physiological space is innate to us, it displays too few resemblances to geometrical space to allow us to see in it sufficient basis for a developed *a priori* geometry (in the Kantian sense)".[8] And further: "Geometrical space is of the same nature everywhere and in all directions; it is boundless and, in Riemann's sense, infinite. Visual space is bounded and finite and what is more its extension is different in different directions, as a glance at the flattened 'vault of heaven' teaches us".[9]

After establishing that human vision does not operate in accordance with a geometrical, Euclidean construction of space, Florensky tackles the problem of how we actually see space, introducing the key notion of the beholder and his/her specific vision. So, the second prerequisite of linear perspective for Florensky

---

[7] Pavel Florensky 'Obratnaia perspektiva' (Reverse Perspective) in Pavel Florensky, *Khristianstvo i kul'tura* (Christianity and Culture), (ed.) A. Filolenko, (Moscow, 2001), p. 88; Pavel Florensky, 'Reverse Perspective' in Pavel Florensky, *Beyond Vision: Essays on the Perception of Art*, (ed.) N. Misler, (London, 2002), p. 265.

[8] Ibid., p. 89; p. 267; The original source by Mach has been lost.

[9] Ibid., p. 89; p. 267; From Ernst Makh, *Analiz oshchushchenii*, (Moscow, 1908), p. 354; Ernst Mach, *The Analysis of Sensations and the Relation of the Psychical to the Physical*, (New York, 1959), p. 181.

is the point of view of the beholder. The transcendental subject thinks that "out of Euclid's absolutely equal points in an infinite space", there exists "a single, *exclusive*, so to speak, monarchical point of particular value, its defining feature being that this point is occupied by the artist himself or more precisely, by his right eye – the optical centre of his right eye. This position is declared to be the centre of the world".[10] Florensky reveals the falsity of that position and rather ironically comments: "There is not a single person, in his right mind, who thinks that his point of view is the only one and who does not accept every place, every point of view as something of value, as giving a special aspect of the world, that does not exclude other aspects but affirms them".[11]

Thirdly, linear perspective presupposes one point of view to the extent of ignoring the double view produced by both eyes.[12] The subject, according to Florensky, imagines himself "as monocular as the Cyclops",[13] because his second eye, rivalling the first, disrupts the uniqueness of vision. The fact that we see with two eyes leads to there being at least two points of view. Here, Florensky draws on a subject, already familiar from the nineteenth century German physicist Hermann von Helmholtz (1821–1894) and namely, the distinction that Helmholtz makes between uniocular and binocular vision in his treatise on physiological optics.[14] Florensky reports of an experiment he conducted to compare the two views produced by each eye respectively. Without going into the details of mathematical figures, suffice is to mention the outcome. It turned out that the difference between the two views was not as small as it might appear at first glance. In fact, the visual image becomes synthetic, binocular and in any case represents a "psychological synthesis, but it can in no way be likened to a monocular, single-lens photography on the retina".[15]

Fourthly, a further prerequisite of linear perspective is the fixed position of the beholder. As Florensky remarks, the subject is thought of "as forever inseparably

[10] Ibid., p. 84; p. 262.

[11] Ibid., p. 90; p. 267.

[12] As could be noticed, Florensky's argument here is basically the same as Panofsky's in *Perspective as a Symbolic Form*. With Panofsky, this forms the first assumption of linear perspective which he attacks. It was an issue, part of the academic discourse at the time. The second assumption of linear perspective, according to Panofsky, is that the flat plane can give an adequate reproduction of our curved "visual image".

[13] Florensky, 'Obratnaia perspektiva' (Reverse Perspective), p. 84; Florensky, 'Reverse Perspective', p. 262.

[14] See Hermann von Helmholtz, *Helmholtz's Treatise on Physiological Optics*, (Rochester, 1925, first in German in 1896), 3 vols, especially vol.3. On Helmholtz and the problem mentioned here, see Marius Tscherning, *Physiological Optics, Dioptrics of the Eye, Functions of the Retina, Ocular Movements, and Binocular Vision*, (Philadelphia, 1904, first in French in 1898), especially p. 216.

[15] Florensky, 'Obratnaia perspektiva' (Reverse Perspective), p. 90; Florensky, 'Reverse Perspective', pp. 267–8.

chained to his throne".[16] That is a false assumption as well, since in reality there is inevitably a constant body movement – of the eyes, head, and so forth. In relation to the movement of the eyes, Florensky draws attention to the scientific fact of saccadic or rapid eye movements discovered and named by Émile Javal at the end of the nineteenth century. In the 1960s it was discovered that this phenomenon was typical even of the state of sleep. The notion of saccadic movements is included in the conception of the "scanning" of pictures in contemporary research[17] and is one of themes which makes Florensky sound very modern.

In the fifth place, not just the beholder but the whole world is conceived "as completely static and wholly immutable"[18] in linear perspective. In reality things constantly move and change. The viewer observes different sides of the objects and never only one. Just as the painter does not contemplate in immobility, the object he contemplates is not immobile itself. In that sense again, the painter has to synthesize and form integrals of particular aspects of reality. According to Florensky, "if the artist wishes to depict the perception he receives when both he and the object are mutually moving, then he must summarize his impressions while in motion".[19] Expressed in mathematical terms, "a single perception halts the process, provides its differential, while a general impression integrates these differentials".[20]

Finally, linear perspective excludes all psycho-physiological processes such as memory. Thus vision becomes "an external-mechanical process".[21] What actually occurs in vision is that "around the dust motes" of what is observed in a sensuous way "the psychic content of the artist's personality crystallizes".[22] Put in simpler terms, movement is not confined just to the outer sphere – a moving subject facing a moving world. There is a lot of internal mental movement going on, as a result of which memories and other phenomena are layered out over the perceived image. In that sense, vision is a highly complicated psychological and physiological process. To really see an object entails the need to successively translate its image onto the retina. In Florensky's words, "the visual image is not presented to the consciousness as something without work and effort, but is constructed, pieced together from fragments successively, sewn one to the other".[23] In discussing this aspect of vision, Florensky is also reacting to a theme, very much popularized by

16 Ibid., p. 84; p. 262.

17 See, for example, C. Gandelman, 'The 'Scanning' of Pictures', *Communication and Cognition*, 19/1 (1986): 3–24.

18 Florensky, 'Obratnaia perspektiva' (Reverse Perspective), p. 85; Florensky, 'Reverse Perspective', p. 263.

19 Ibid., p. 92; p. 269.

20 Ibid., p. 93; p. 270.

21 Ibid., p. 85; p. 263.

22 Ibid., p. 93; p. 270.

23 Ibid., 93; p. 270.

Adolf Hildebrand (1847–1921) in his essay 'The Problem of Time in the Fine Arts' (1893).[24]

As was mentioned, the six premises of linear perspective do not apply in the case of reverse perspective. The space in the icon *does not* follow Euclidean laws (first premise). There is *no* absolute point of view of the beholder (second premise). The various viewpoints represent different aspects of the object and of reality at large. Further, reverse perspective takes into consideration the double view produced by the two eyes, a fact disregarded by linear perspective (third premise). A further characteristic of the beholder is that he/she is no longer assumed to occupy a fixed position (fourth premise).

With reverse perspective, the typical situation is that some movement on part of the painter is presupposed while reality can be less mobile or completely still (fifth premise). We are dealing here "with the unmoving monumentality and ontological massiveness of the world, activated by the cognizing spirit".[25] The various visual perceptions produced by the different points of view are then psychologically synthesized. In that way, the pictorial image becomes an integral of varied –and always double/binocular –images. It is this procedure that brings about the phenomenon of simultaneous planes, discussed in the last chapter of this book. "The spiritual-synthetic nature of visual images"[26] is a key notion in Florensky's analysis and extremely influential in Russian art criticism.

Finally, reverse perspective, unlike linear perspective takes account of an inner, psychological factor (centring on the process of memory) which influences perception (sixth premise).

We can safely maintain that Florensky's whole discussion of the theoretical premises of linear perspective, as outlined briefly above, is conducted from the point of view of a narrow definition of mathematically constructed pictorial space, as the one put forward by Leon Battista Alberti during the Renaissance.[27] Florensky himself was well aware that deviations from strict perspectival rules were, in fact, common practice in the Renaissance and later.[28] At the same time, he

---

24 See Adolf Hildebrand, 'The Problem of Form in the Fine Arts' in Harry Mallgrave and Eleftherios Ikonomou, (eds), *Empathy, Form and Space: Problems in German Aesthetics 1873–1893*, (Santa Monica, Cal., 1994), pp. 227–81.

25 Florensky, 'Obratnaia perspektiva' (Reverse Perspective), p. 92; Florensky, 'Reverse Perspective', p. 269.

26 Ibid., p. 91; p. 268.

27 The way in which linear perspective can implicate time has been discussed by several authors (see Note 53). Martin Kemp shows that in the sixteenth century, paintings frequently implied a moving viewer, rather than a viewed fixed in space, as Alberti had suggested (Martin Kemp, 'From Different Points of View: Correggio, Copernicus, and the Mobile Observer', in Lars Jones and Louisa Matthew (eds), *Coming About ... A Festschrift for John Shearman,* (Cambridge, MA, 2001).

28 See Florensky's discussion of the conscious perspectival deviations in some great Renaissance masterpieces, as Leonardo's *Last Supper*, Raphael's *School of Athens*, etc.

sees the theory that informs Renaissance perspective as growing out of a certain, rationalistic worldview that he set out to discredit by undermining its claims for providing a true, scientific model of the world.

Another issue arises, when some of Florensky's claims, which may be true in abstract science, do not hold ground in terms of the actual perception of pictures. Complete immobility is an impossibility, as Florensky points out, though in practice I do not see why we should not imagine such a fixed position in everyday terms. Another reason due to which we do not perceive just one aspect but many, according to Florensky, is that the objects themselves are characterised by mobility and change. In practice, it is quite possible, as long as the viewer is in a fixed position, that he/she sees a single aspect of an object and it remains stable in its relationship to other objects.

A completely different matter is the fact that some of the specifically scientific research, that Florensky brings forth, is now outdated. For our immediate purposes it is unnecessary to address the problem of the scientific truths and shortcomings of the six conditions discussed by Florensky. What is important is the major thrust of the argumentation. The implication is that while the scientific foundations of linear perspective prove unstable, reverse perspective remains closer to the way vision functions. The second part of the proposition is highly problematic, as we certainly do not "see" in reverse perspective, though it is possible that it may correspond in some way to aspects of the visual process. However, the heart of the problem was to defend icon art as truer to vision in the light of contemporary ideas. The approach itself is rather obsolete as it is in essence a repetition of the Renaissance claim that the new art has the characteristics and therefore the status of science. Whatever the value of a Michelangelo, a Leonardo or a Byzantine icon, a fresco by Rublev, a modern art-lover seeks for it elsewhere than in an artistic adherence to optical laws.

For current research on reverse perspective, Florensky's account is of interest for two reasons. Firstly, it has exerted decisive influence on subsequent Russian criticism on the subject. It could be said that Florensky provided the basic orientation from which further studies have proceeded. Secondly, while Florensky is mainly interpreted within a Russian and Eastern Orthodox context, his work is also very expressive of the general intellectual problems of the early twentieth century. His approach is determined by research at his time. The theories in non-Euclidean geometries, the ones in the field of vision were part of the same atmosphere that gave rise to studies on reverse perspective in German.[29] On the whole, it seems that our current notions of reverse perspective were first formulated by German authors and from there were taken on and elaborated, largely through Florensky,

---

(Florensky, 'Obratnaia perspektiva' (Reverse Perspective), pp. 63–4; Florensky, 'Reverse Perspective', pp. 229–31).

[29] For the best account of the intellectual impact of non-Euclidean geometries, see Linda Henderson, *The Fourth Dimension and Non-Euclidean Geometry in Modern Art*, (Princeton, 1983).

by Russians. The whole scheme is best to be considered within the context of a general movement of ideas at the beginning of the century. The problem here is whether the ideas of reverse perspective are still as valid and adequate.

Second Stage: Zhegin

Lev Zhegin (1892–1969)'s *Iazik zhivopisnogo proizvedeniia: uslovnost' drevnego iskusstva* (The Language of the Work of Art: Conventionality of Ancient Art) (1970) is the result of forty years of work. The Russian writer's interest in ancient systems of spatial construction was first excited in the 1920s. Florensky's influence at the time must have been considerable, especially as the two knew each other and moved in the same intellectual circles. Indeed, that Florensky's ideas on "reverse perspective" provide the starting ground for Zhegin is obvious throughout the latter's work.[30] It is also natural, that Zhegin, who was himself an artist of no small merit,[31] should take the turn he did and namely provide an actual visual analysis of Florensky's more general theory.

Zhegin's book seems to have attracted relatively little attention outside the Russian-speaking world, even though there is a German translation of it.[32] Zhegin's is perhaps the most comprehensive scheme of the way reverse perspective functions in optical and geometric terms. The author offers a visual analysis of ancient systems of perspective and particularly of reverse perspective. It is also practically useful as a guide for deciphering ancient images and noticing how they function. The work is especially addressed to a spectator accustomed to images treated by linear perspective. Such a spectator needs to make a conscious effort to become comfortable with depictions formed according to different systems. And this is where Zhegin provides a helping hand.

A foremost concern has been, just as in Florensky's earlier essay, to dispel the illusion that perspectival deformations are the result of the artist's inability to master mathematical perspective. The consistency of the deviations suggests that they are not due to technical or artistic failure on part of the artist.

"Deviations" can be grouped according to their character and, in particular, according to the position of the vanishing point/points in relation to the horizon. While with linear perspective the vanishing point tends to lie on the level of

---

30 Zhegin himself acknowledges this in his admiring memoir of Florensky (Lev Zhegin, 'Vozpomenaniia o P. A. Florenskom' (Memories of P. A. Florensky) in Konstantin Isupov, (ed.), *P. A. Florensky: Pro et Contra* (St. Petersburg, 1996), pp. 162–73.

31 Zhegin's works are in the Tretyalkov Gallery in Moscow and in other Russian galleries.

32 Lev Shegin, *Die Sprache des Bildes*, (Dresden, 1982). I thank Boris Uspensky for drawing my attention to this. Egon Sendler mentions Zhegin in his book *Icon: Image of the Invisible*, (Redondo Beach, Cal., 1988, first in French in 1981), but does not seem to do full justice to Zhegin's work.

the horizon, Zhegin distinguishes between two other positions that define two variations of reverse perspective. In cases when the vanishing points are above the horizon level, we can talk of hidden forms of reverse perspective. If the vanishing points are below the horizon level and even below the basis of the objects, the so-called obvious forms of reverse perspective occur, which are the ones of interest for the on-going discussion (Fig. 2.1).

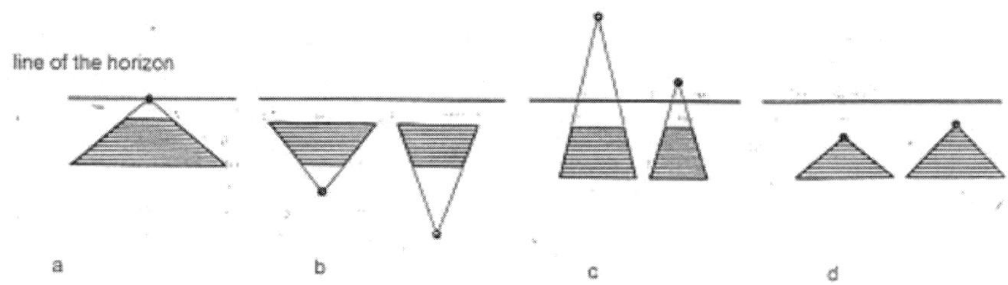

Fig. 2.1 Different types of perspective forms according to the position of the vanishing point in relation to the line of the horizon (Fig.1 in Zhegin):

a. Linear perspective: the vanishing point is on the horizon line
b. Obvious forms of reverse perspective: the vanishing point is below the basis of the object
c. Hidden forms of reverse perspective: the vanishing point is above the line of the horizon
d. Intensely converging perspective: the vanishing point is below the horizon and above the object

In either case, whether above or below the horizon level, a single picture has more than one vanishing points. Actually, the multiplicity of vanishing points is a typical feature of the images constructed in reverse perspective. With the obvious forms of reverse perspective the vanishing point is below the basis of the object. Thus the multiplicity of vanishing points is due to the tendency of each object to have its own vanishing point. In practice, the number of vanishing points becomes equal to the number of objects. With the hidden forms of reverse perspective and under the conditions of greater spatial depth, as in Renaissance art, the multiplicity of vanishing points is reduced (though, Zhegin finds no less than ten in Veronese's *The Wedding at Cana*). Above all, it no longer corresponds to the number of objects. The variety of potentially available viewing positions has decreased and so has the unfolding of the images, the variety of the aspects of the same image.

The multiplicity of vanishing points requires multiple viewpoints and that, in its stead, gives rise to a dynamic view on part of the beholder. The dynamic view suggests a change of position and the simultaneous co-existence of different aspects of the same image as a result, in a manner similar as the one claimed for Analytical Cubism (see the beginning of Chapter 4). Consequently, the need arises to combine the various positions and aspects into one synthetic image. Thus the image treated by reverse perspective does not represent one single aspect of the figure (as with linear perspective), but a synthesis of several aspects. The very obvious and frequently observed perspective deformations are the result of this process. Several views that cannot be seen at the same moment overlap in a single representation. To achieve such a view would have required movement in time of the beholder, the image, or both.

In light of this process of seeing and representation, how can we make sense of an image like the one in Figure 2.2?

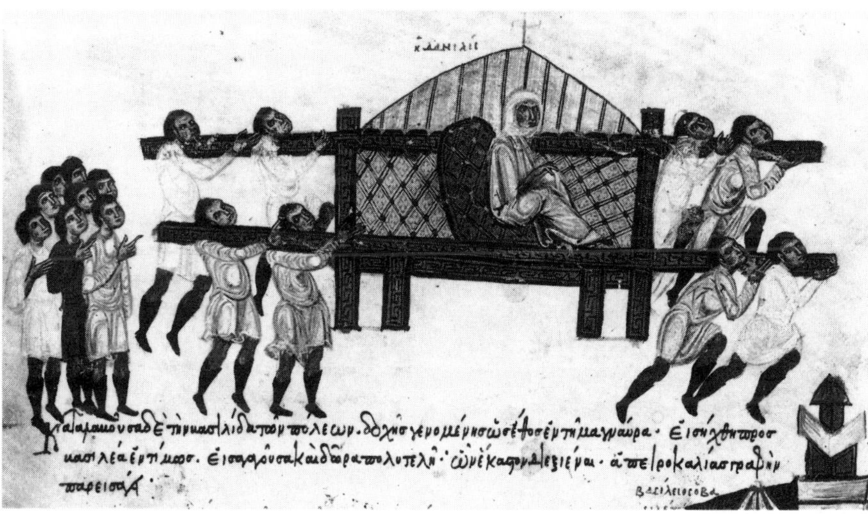

Fig. 2.2 Madrid Chronicle of John Skylitzes, *The Landowner Danielies Carried on a Litter,* 12th c. (?), (cat. no. 338), fol. 102 r (a) Biblioteca Nacional, Madrid

What seems at first glance to be a figure on the background of something of a quadrangular shape turns out to be a figure carried on a litter. The upper two sets of figures are not hanging onto a horizontally placed pole but are actually carrying the pole. The litter is represented as if seen simultaneously from the side and from a bird's view. It is a complicated image to decipher for anyone who attempts to read it according to the tenets of standard perspective. The key to it lies in the effects produced by "reverse perspective".

The same phenomenon occurs, for example, in numerous representations of the Bible, in which the cover of the book is depicted frontally but at the same time we can see two of the sides or in various images of *The Birth of the Virgin* which strike us with the eccentric shape of the cot until we discover that it is a simple quadrangle of which three sides are given simultaneously.

This phenomenon becomes especially explicit in the depiction of architecture, as in Figure 2.3.

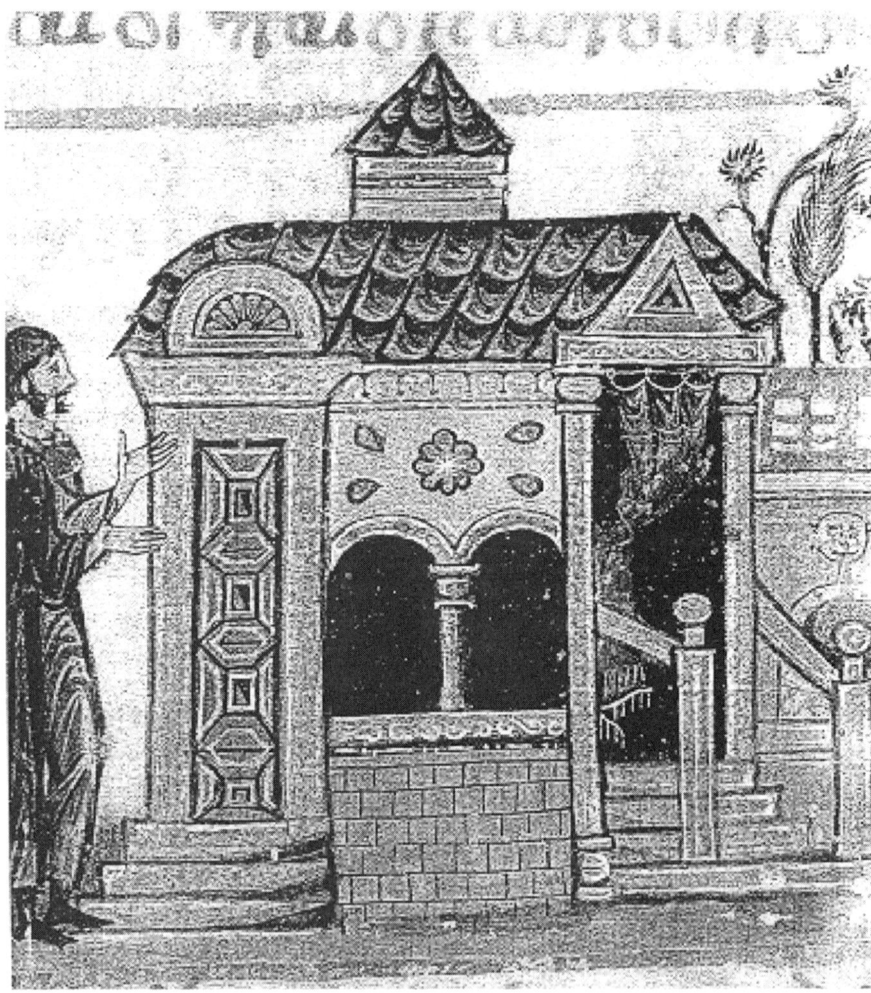

Fig. 2.3 *The House of Eglon, King of Moab*, Vatican Octateuch, Biblioteca Vaticana,12th c.; MS Vat.gr.746

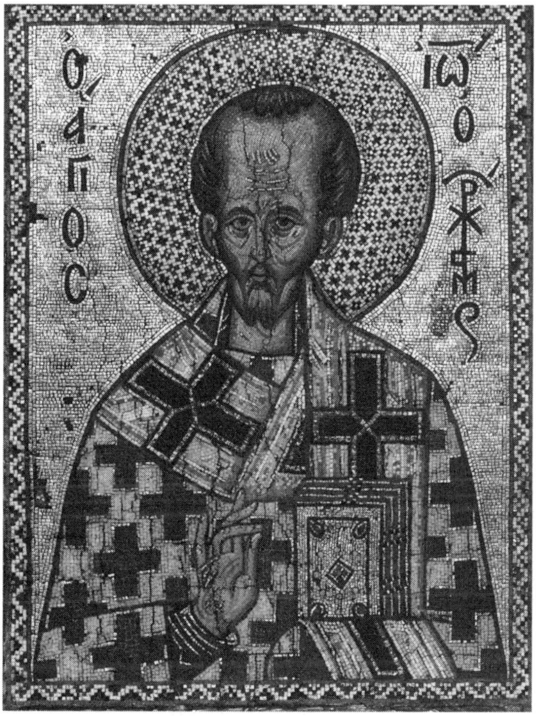

Fig. 2.4 Representation of the face in the system of concavity: *St. John Chrysostom*, Palaeologan mosaic icon, Dumbarton Oaks Collection, number 352

There are two staircases leading to two entrances which in fact are on two lateral sides. The one on the right has the banisters placed at an oblique angle, but the entrance itself gives the false impression of a frontal view. This effect is again due to the perspective treatment.

Of course, this phenomenon is not confined only to geometrically simple forms. The principle of simultaneous depiction of different planes is extended to more complicated forms, as for example, those of the face. According to Zhegin, it lies at the heart of a typical facial type with a disproportionately wide forehead. The total shape of the face becomes triangular, an impression is enhanced by the pointed beard. It is an extremely common facial type and has often been remarked upon by various scholars. What usually evades attention is that the triangular form is an effect of the adding up of planes in the upper part of the face, where aspects of the profile view are added to the frontal view. Explaining the configuration as a synthesis of planes provides an alternative to the sort of interpretations that argue: "The ears are always shown uncovered as they are created to listen to and hear the commandments of God".[33] In fact, the ears of the figures in icons are most often represented as they would look as seen in full view which is then added to the frontal position of the face. The same explanation lies at the heart of the frequently met protuberance of the forehead (Fig. 2.4).

The popular facial type with a wide forehead sharply narrowing towards the chin can be interpreted in the light of the above said. Panofsky attributes "the

[33] Stuart Robinson, *Images of Byzantium* (London, 1996), p. 23.

unnatural heightening of the cranium"[34] to the canon of proportions in Byzantine and Byzantining art, where a module system was applied (the unit being the length of the nose).[35] Zhegin provides a convincing, alternative explanation to a facial type with a long history in Byzantine and Byzantining art.

Thus, the following could be accepted as a working definition of reverse perspective – it is "directly connected with the dynamics of the viewing position: the form of reverse perspective is the result of the summarizing of the viewer's perception under the conditions of a multiplicity of viewpoints, that are themselves the result of the dynamics of the viewing position".[36]

In the process of summarizing the viewing positions, the sides of the objects unfold (Fig. 2.5).

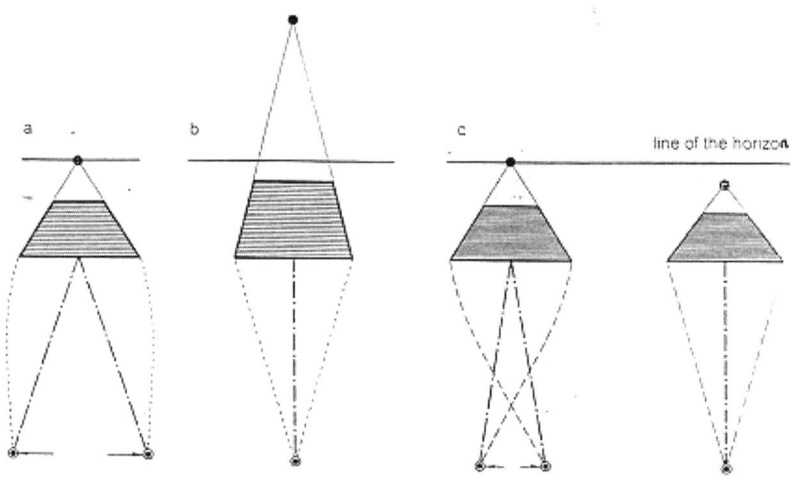

Fig. 2.5 Reason for the perspective deformations: the dynamics of the viewing position and its summarizing (Fig. 2 in Zhegin):

a. Split up viewing model
b. Summarizing the model
c. Intensely converging perspective

---

[34] Erwin Panofsky, 'History of the Theory of Human Proportions as a Reflection on the History of Styles' in his *Meaning in the Visual Arts* (Harmondsworth, 1970), p. 109.

[35] Panofsky's hypothesis of the unit system has been disproved in June Winfield and David Winfield, *Proportion and Structure of the Human Figure in Byzantine Wall-painting and Mosaic* (Oxford, 1982).

[36] Lev Zhegin, *Iazik zhivopisnogo proizvedeniia: uslovnost' drevnego iskusstva* (The Language of the Work of Art: Conventionality of Ancient Art), (Moscow, 1970), p. 42.

As a result, the volumetric forms become flat, while the flat ones become concave. Figure 2.6 shows such a concavity of the image, as a result of the summarizing of the split viewing model.

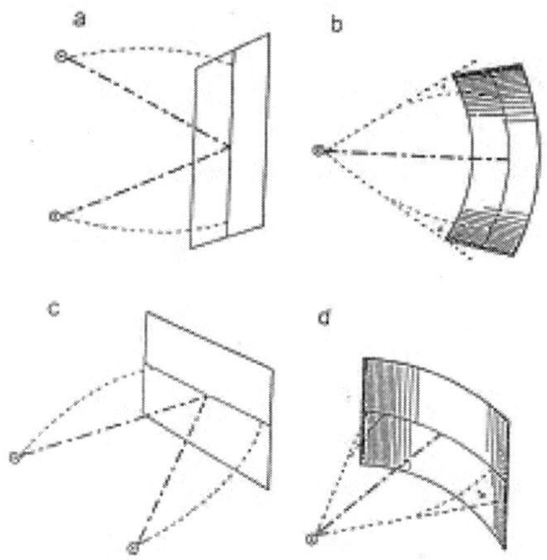

Fig. 2.6 The summarizing of the split viewing model creates the system of concavity (Fig. 3 in Zhegin):

a. Dynamic position (the ray is curved). Split position along the vertical
b. In the process of subjective visual straightening of the ray there occurs a summarizing of the position; it adheres to one fixed point of view – the image curves
c,d. The same phenomenon in the case of a split position along the horizontalactual form of the objects the form, after transformation in the system of concavity

The complicated and dynamic viewing position most often leads to seeing the object both from above and from the sides. If the object is in the air an aspect from below is often given. In other words, the image unfolds and different aspects are added up together. As Zhegin puts it, "the dynamics of our position is transferred onto the object, the form becomes dynamic – it becomes concave".[37]

The author further offers a substantial enumeration of certain typical "deformations" that forms undergo in the system of concavity (Fig. 2.7).

[37] Ibid., p. 43.

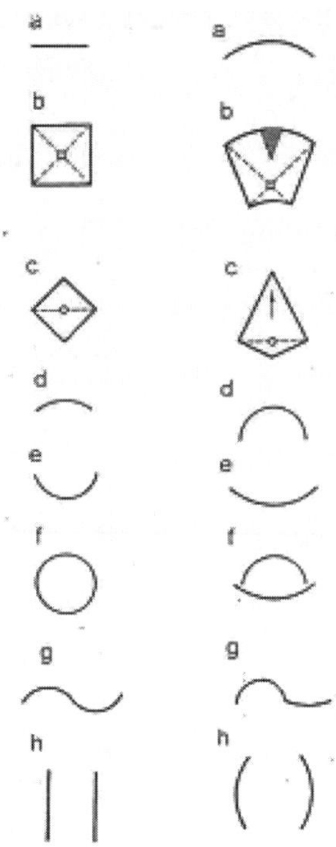

Fig. 2.7 Deformations of the forms in the system of concavity:

a. The straight line becomes curved
b. The front and the back sides of a rectangular form become curved, while the lateral sides tend to divorce. The centre moves towards the viewer
c. The far angle of a rectangular form becomes more narrow and moves away from the viewer, while the front angle becomes wider. The centre moves towards the viewer
d. The concave line becomes even more concave
e. The convex line tends to straighten
f. The back part of a circle curves even more strongly, while the front tends to straighten. The form of the circle breaks
g. The concave part of a wave-like line becomes even more concave, while the convex tends to straighten. In the place of the transition of the one form to the other, there occurs a break within the form
h. The two parallel verticals become barrel-like in form

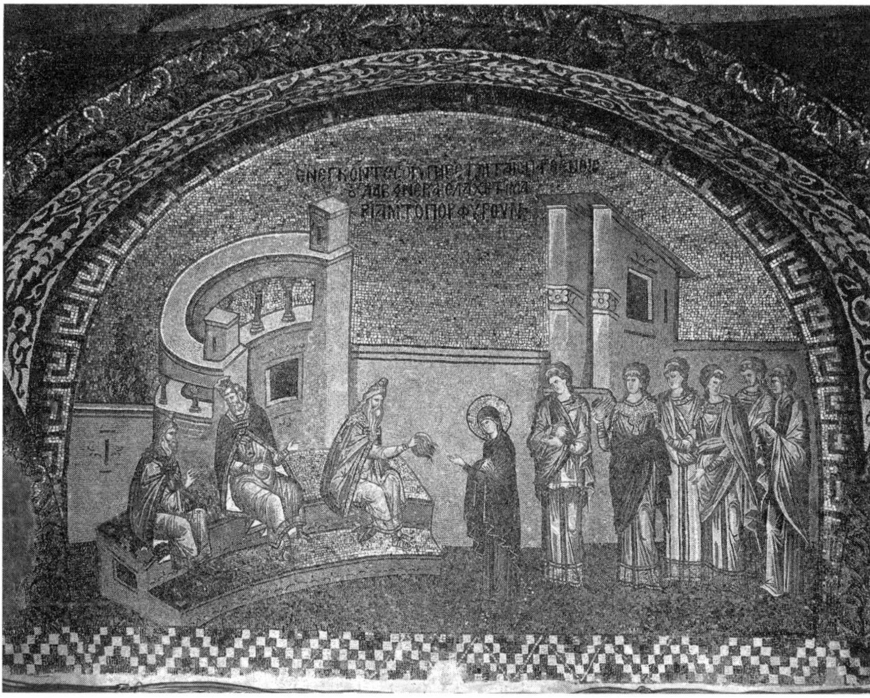

Fig. 2.8 "Mary Receiving the Purple", fragment, Kariye Camii, Constantinople, c. 1304, inner narthex, bay 3, West, wall lunette.

Figure 2.8 is a particular pictorial expression of Figure 2.7a and demonstrates how straight lines become concave. The original form of the architectural structure should be rectangular rather than curved.

As a result of the tension that occurs in the process when a straight line becomes concave (Fig. 2.7a) sometimes a partial or complete breaking up of the form appears in the centre of the concave line. In Fig.2.9 the back of the throne is partially split. This phenomenon affects not only simple geometrical forms but also the treatment of the human face, as can be seen in a relief (Fig.2.10), in which the forehead of the figure is vertically split up in the centre.

When considering some of the deformations in images treated by reverse perspective Zhegin claims that the "barrel-like" form is "one of the most characteristic deformations".[38] Rectangular forms in reverse perspective appear as if drawn on a concave surface and so are realised in what Zhegin calls the "barrel-like form" (Fig. 2.7h). This form is the result of the process of summarizing of the viewing position, split in two mutually perpendicular directions – the vertical and

[38] Ibid., p. 46.

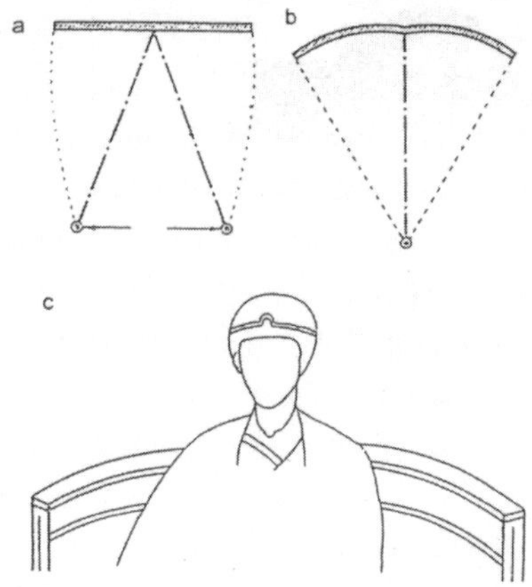

Fig. 2.9 Partial breaking up of the form in the centre (Fig.6 in Zhegin)

a. Split up model
b. Summarized model
c. As a result of the summarizing of the split-up model, the throne becomes double-centred

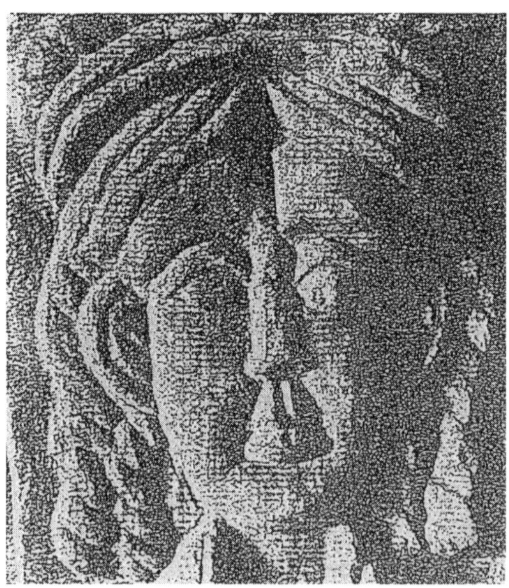

Fig. 2.10 Visual effect in the form of a partial breaking up of the form (here – in the centre of the forehead) as a result of the summarizing of the split-up model: (Pl. Ve from Zhegin), 1230–1234; fragment from a relief, St. George's Church, Yurievo-Polsko, Russia

the horizontal. A good example of such deformation is the treatment of the back of the ancient Russian throne or chair (Fig. 2.11).

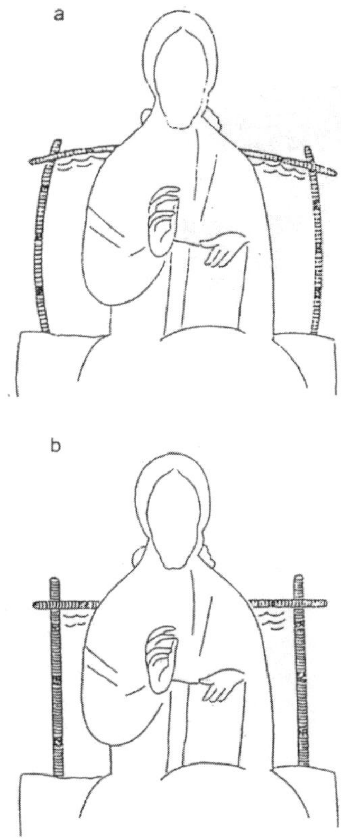

Fig. 2.11 Form of the throne (Fig.8 in Zhegin)

(a) in a "barrel-likc deformation"
(b) actual form of the throne

Other perspectival deformations include what Zhegin calls the move "from us" and the move "towards us", as well as the move "to the side". The first happens in the process of unfolding the sides of the object, as a result of which the image is flattened. The decrease in spatial depth has as one of its consequences the enlargement of the more distant part of the image – the move "from us". At the same time, the objects change their logical position in the direction to the beholder – the move "towards us" (Fig.2.12).

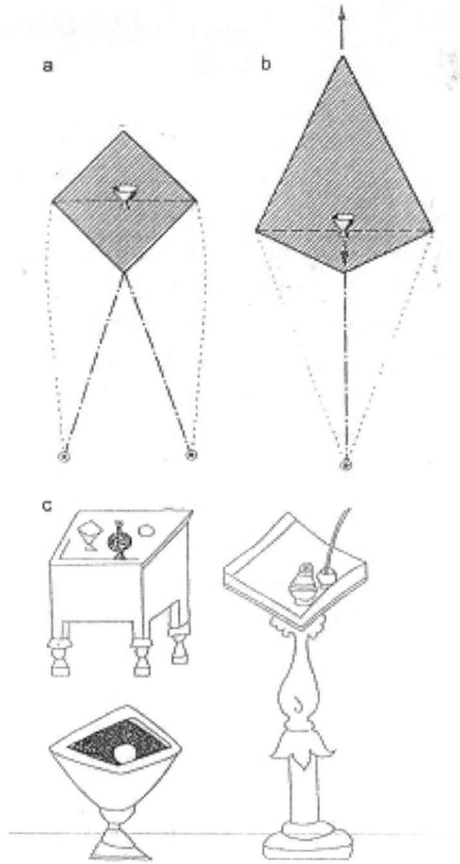

Fig. 2.12 Objects situated in the centre make a move towards the viewer:

a. Split up viewing model
b. Summarized viewing model. The far end of the object moves towards the viewer
c. Objects having moved towards the viewer

This is what frequently happens in representations of the chalice – as in images of *The Last Supper* – the position of which has moved towards the lower end of the table and thus in the direction of the viewer. The deformation affected as a move "to the side" is the result of the summarizing of the horizontally split (Fig. 2.13a) and vertically split position (Fig. 2.13b).

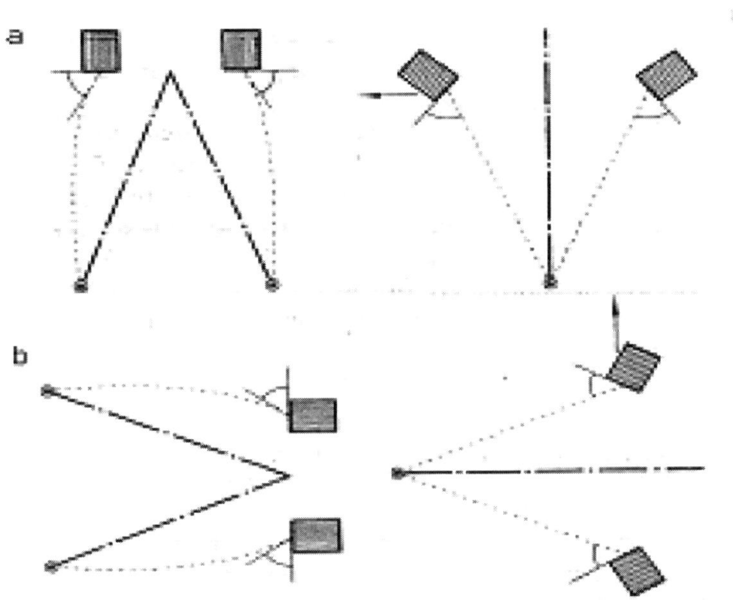

Fig. 2.13 Summarizing the model causes "moves" in the representation:

a. Along the horizontal
b. Along the vertical

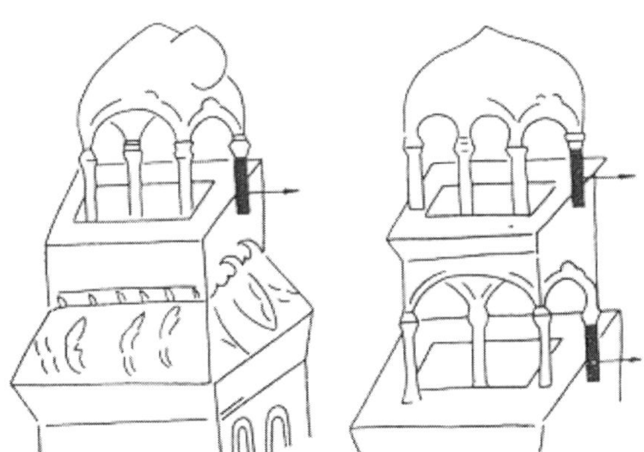

Fig. 2.14 Visual "move" to the side: From the icon of the A*nnunciation, Novgorod School*, 15th c.

Objects and parts of objects hanging in the air without their natural support are explained in this context. Some of the columns, particularly those closer to the edge lose their stability and appear as hanging from the building (Fig. 2.14).

In the system of reverse perspective the "wave-like" form undergoes a specific transformation, as can be seen on Fig. 2.7g. A concave form becomes even more so (Fig. 2.7d), while a convex one shows a tendency to straighten, sometimes becoming fully straight (Fig. 2.7e). Combining the transformation that a concave form undergoes and the one of a convex form, an unrealistic form as that of the cup on Fig. 2.15 is realised. Instead of a gradual transition from concave to convex an abrupt partial breaking up of the form appears. This phenomenon is especially conspicuous with simple geometrical forms. It occurs in more complex forms as well.

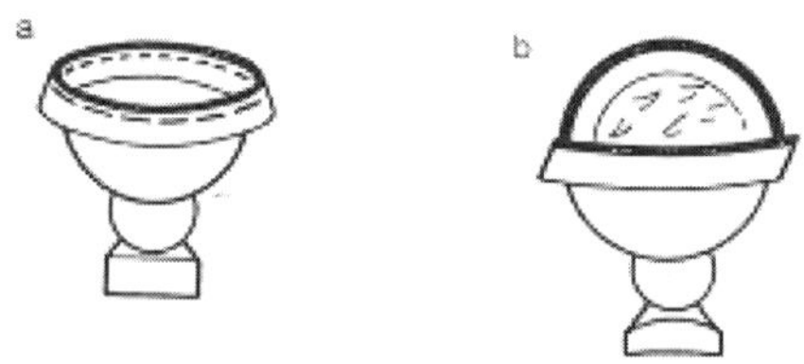

Fig. 2.15 Deformation of the circle in the system of concavity:

a. Actual form of the cup as seen from a distance
b. Representation of the cup in icon art

While with linear perspective there is an ever-widening field of vision, with reverse perspective the field of vision is narrowed. This, according to Zhegin, is a characteristic of the more ancient types of art and is possible only under the conditions of a multiple viewpoint.[39] The more viewing positions the less scope each of them covers. Zhegin's analogy is with the image produced by a telescope – a strong lens increases the size of the images while, simultaneously, the field of vision narrows reducing the number of objects under view. As can be seen in Fig. 2.16 "to realise the forms of reverse perspective there should be, above all, a sense of isolation from the neighbouring positions.

---

[39] Ibid., p. 52.

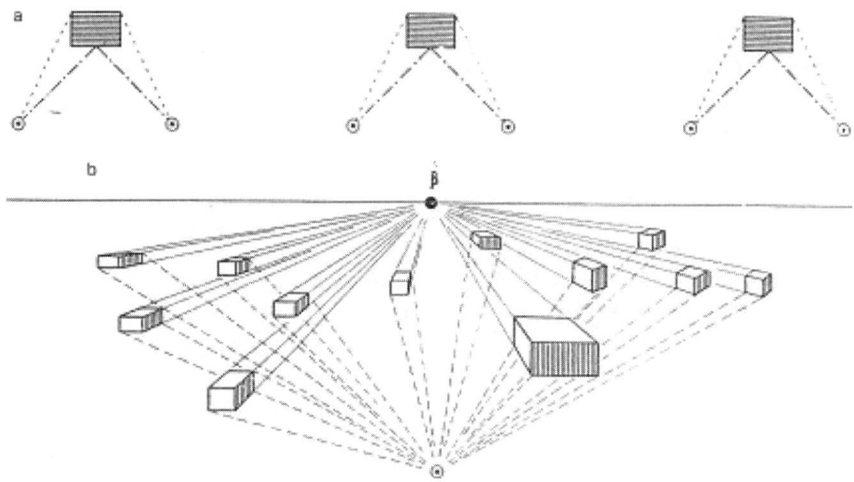

Fig. 2.16 Visual scope of dynamic and linear space (Fig. 14 in Zhegin):

a. In the system of dynamic space, each object has an independent viewer, whose point of view is dynamic and split up. There can be different degrees of dynamics

b. In the system of linear space, the multitude of objects are taken in by a single point of view, to which a single vanishing point corresponds

The rays of vision should be focused exclusively on the given object – in other words, there should be a 'narrowed' field of vision".[40] Thus we return to what was noted earlier, namely that in reverse perspective there are as many vanishing points as there are objects. The viewpoint corresponds to a vanishing point and, consequently, there are as many points of view as vanishing points, which means that in this perspectival system only one object occupies the field of vision at a time.

The perception of images treated by reverse perspective requires a method of "slow reading" – each object is perceived on its own. To compensate for the multiplicity of points of view, there is a tendency for reducing the number of objects represented. Zhegin gives the example of a couple of leaves suggesting the tree's crown and a few people gathered together symbolizing a crowd (often there is only an outline of heads and shoulders to suggest a multitude).

---

[40] Ibid.

A further peculiarity of ancient art is the tendency towards the levelling of scale, a phenomenon Zhegin sees as disappearing only with the nineteenth century. Before that, levelling was typical especially of images where the third dimension was suppressed (though Zhegin cites a fresco by Raphael where the camel in front of the walls of Jericho and the wall itself are of the same height). Thus, in the Sinai icon of the Mother of God (Fig. 2.17) the heads of the standing and the seated figures are on the same level.

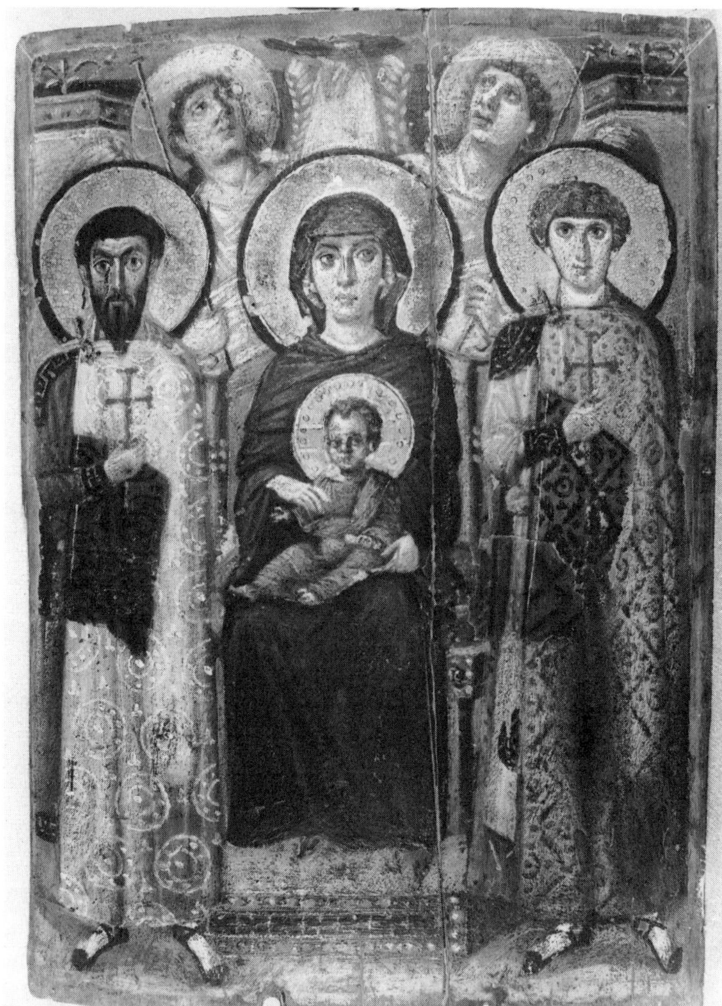

Fig. 2.17 Levelling of sizes: *The Enthroned Mother of God between Sts. Theodore and George*, tempera on wood, Monastery of St. Catherine, Mount Sinai

Finally, Zhegin points out the dominant viewing position as another characteristic of ancient art. Within the multiplicity of viewpoints there is a consistent organisation of the whole, a compositional unity where one point of view dominates over the rest. The dominant viewing position means that not only each object perceived from a particular viewpoint is subjected to the system of concavity but the composition as a whole is concave. As Zhegin notices the dominant position is similar in its structure to the secondary positions.[41] The many-sided range of vision required for the perception of separate figures is activated in respect to the picture as a whole, as well.

Let us summarize the main points of Zhegin's exposition on the transformation of images in the system of reverse perspective. The problem is seen predominantly in terms of the vanishing point/vanishing points – a terminology and concept borrowed from the theory of linear perspective. With obvious forms of reverse perspective, the vanishing point lies below the horizon level and below the basis of the object. Due to the tendency of each object to have its own vanishing point, the phenomenon of the multiple vanishing points becomes a major characteristic of the discussed perspective system. Correspondingly, a multiplicity of viewpoints is presupposed that gives rise to a dynamic view. The resulting images require a process of synthesis. The unfolding of the synthetic image of reverse perspective reveals different aspects of the image which are represented simultaneously. The volumetric forms of these images flatten, while the flat ones become concave. A typical phenomenon is a partial breaking up of the concave form which sometimes occurs.

Particular perspective deformations concern the "barrel-like" form, the move of the objects in images "towards us", while the distant parts of the image move away "from us", the "move to the side", and the specific treatment of the "wave-like" form. These typical deformations happen in a narrowed field of vision. Their perception requires a method of "slow reading" which goes hand in hand with the reduced number of objects. Sometimes, there is a tendency towards the levelling of figures. The objects treated by reverse perspective are seen each from its own viewpoint, but the image as a whole is perceived from one dominant viewing position.

Two main motifs can be identified in Zhegin's argument and both can be traced back to Florensky. The first is that the "reverse perspectival" image is the result of a summarized and synthesized view of several aspects of the represented object. This thesis is a major thread that runs through Florensky's 'Reverse Perspective' and I believe it constitutes a true insight into the principle of the spatial construction of icon art. The second motif of importance in Zhegin's work is the claim that every object in "reverse perspective" has its own vanishing point. This is a highly problematic observation and I believe, ultimately untrue.

The term "vanishing point" describes a very particular phenomenon – as a result of precise, and often laborious measurement, all parallel lines within the

[41] Ibid., p. 56.

pictorial composition converge into a single point. There is a possibility, of course, for more than one vanishing point, as was mentioned by Zhegin himself and even a vanishing point (or points) which are indicated by the orthogonals to lie outside the picture surface. In any case, whether one or more, such vanishing points are the result of specific measurements.

In the case of "reverse perspective", on the other hand, there is no evidence that icon-painters had recourse to mathematically correct systems of measurement to enable them to represent vanishing point systems, whether single or multiple ones. It is very unlikely that, if we make the necessary measurements, Zhegin's contention would be confirmed. It is much more likely the lines of the objects in "reverse perspective" very often *intersect*, more or less at random and not in a systematic manner, rather than converge in a single point.

The question is why would Zhegin put forward such a claim? He definitely found some support in Florensky's essay, but even in this earlier work there is a greater caution in the formulation. At one point, Florensky mentions that perspective can be "linear or reverse, with one or many centres"[42] and so gives good grounds for Zhegin's "vanishing point" versus "vanishing points" in linear and "reverse perspective" respectively. However, it deserves mention that Florensky could have used the more technical term, but he chose not to. Rather, he uses the much looser "centre" which could be taken to imply the tendency of the convergence of parallel lines in "reverse perspective" in front of the picture surface, but not necessarily their actual, strict coming together into one point. In any case, it is Zhegin who put the term into circulation for icon art – a term, originally coined to describe a Renaissance artistic practice. To expect parallel lines to converge is a logical outcome of the approach of employing concepts of linear perspective for the explanation of reverse perspective. As mentioned above, I believe this transfer of terminology to be unjustified. With it, however, Zhegin, it could be said, made the second step, after Florensky, in the process that has brought about our present ideas of "reverse perspective"[43]. I have been trying to make clear my conviction that this step was taken in the wrong direction.

Having said that, however, it takes away none of the merit of Zhegin's otherwise illuminating study, which provides a post-medieval audience with a visual key with which to decode images characterized by a non-standard system.

[42] Florensky, 'Obratnaia perspektiva' (Reverse Perspective), p. 75; Florensky, 'Reverse Perspective', p. 254.

[43] None of the writings, devoted to the theory of "reverse perspective", question Zhegin's assumption of the multiple vanishing points. See, for example, Sendler, *The Icon: Image of the Invisible*, p. 34.

Third Stage: Boris Uspensky

The third step in the development of the theory of "reverse perspective", after Florensky's and Zhegin's, was taken by the contemporary Russian scholar Boris Uspensky.[44]

Uspensky is of interest, firstly, for his role in popularizing the research done on "reverse perspective". His mission in this respect cannot be separated from the remarkable Soviet journal, entitled *Trudy po znakovim sistemam* (Researches on Sign Systems). The editor-in-chief was Iuri Lotman, the well-known specialist on semiotics. Uspensky was a contributor from the very beginning, when the journal was launched in 1965 and part of the editorial team since 1984. The volumes of the Researches provide fascinating reading, with a host of great names, for whom the journal provided a forum for the discussion of new, original and often unorthodox ideas, sometimes voiced for the first time. Florensky's "Reverse Perspective" was published for the first time in the third volume of the journal,[45] the author and the work being introduced in a short preface by Uspensky, together with A.A. Dorogov and Viacheslav Ivanov.[46] Zhegin's first drafts of his study were also first published in the Researches. When they came out in a book form, the introduction again was by Uspensky,[47] who by that time had become deeply involved with the problematics of space in ancient Russian art. Given this background, Uspensky was well qualified to synthesize the Russian tradition of writing on "reverse perspective". He also left his own stamp on this development by pushing into the foreground the motif of the "inner viewer". Thus, Uspensky took what I call the third and, in terms of the Russian theory on the topic, final step. He was partly responding to ideas, hugely popular at the time of the publication of his work *The Semiotics of the Russian Icon* in the fifth volume of *Trudy po znakovim sistemam* (Researches on Sign Systems) (1971),[48] centring on the role of the viewer/reader in the construction of meaning. Ironically, in the context of the theory on "reverse perspective", his argument takes us full circle to Oscar Wulff's main thesis in "Die umgekehrten Perspektive und die Niedersicht" (1907).

---

44 Care should be taken that Boris Uspensky is not confused with Vladimir Uspensky as both write on the subject of the meaning of the icon, though the former's approach is semiotic, while the latter tries to evolve a theology of the icon.

45 See Florensky, 'Obratnaia perspektiva' (Reverse Perspective), *Trudy po znakovim sistemam* (Researches on Sign Systems).

46 Boris Uspensky, A. A. Dorogov, and Viacheslav Ivanov, 'P. A. Florensky i ego statiia "Obratnaia perspektiva" (P. A.Florensky and His Essay 'Reverse Perspective', *Trudy po znakovim sistemam* (Researches on Sign Systems), Tartu, 3 (1967): 378–80.

47 Boris Uspensky, 'K issledovaniu iazika drevnei zhivopisy' (Towards a Study of the Language of Ancient Art) in Zhegin, *Iazik zhivopisnogo proizvedeniia: uslovnost' drevnego iskusstva* (The Language of the Work of Art: Conventionality of Ancient Art), pp. 4–34.

48 See Boris Uspensky, 'Semiotika russkoi ikoni' (The Semiotics of the Russian Icon), *Trudy po znakovim sistemam* (Researches on Sign Systems), 5 (1971): 178–222.

What Wulff had proposed for the first time is intimately connected to the vanishing point construction, as discussed in relation to Zhegin's study. A vanishing point construction implied its counterpart in the viewer. In this sense, if the term is employed in the theory of "reverse perspective" it is imperative that an account is provided for the nature and position of the beholder of "reverse perspectival" images.

Wulff's contention on this point is strikingly simple – the viewer, external to the pictorial composition, is invited to mentally project himself as an internal viewer. Once part of the pictorial space, this internal viewer perceives the representation exactly as the external observer of linear perspective – in respect to his inner standpoint, the parallel lines of each object converge in a vanishing point and as a result, objects look "normal". The neat proposition is based throughout on a mechanical application of notions, worked out in the system of "reverse perspective" – "picture surface" (with a distinction behind and in front of it), "vanishing point", a particular interpretation of "internal-external viewer". I have indicated already several times why I do not accept the whole procedure. I will now concentrate on the last pair of concepts and the role they play, according to Uspensky.

The beholder's position was taken into account both by Florensky and by Zhegin. As we saw, the second, third and fourth premises of Florensky directly relate to this issue, just as Zhegin's "multiple vanishing point" construction. With Uspensky, however, the viewer becomes the main focus of interest. The definition of "reverse perspective" itself depends on the specificity of the implied viewing position. While with linear perspective the two positions – of the artist and the viewer – ideally coincide, this is not the case here. In contrast to the observer in linear perspective, who perceives the image from a position in front of it, the inner viewer of "reverse perspective" is on the other side of the picture surface. To emphasize this distinction, Uspensky uses two different terms – "beholder", who is part of the pictorial space and "viewer", who inhabits his own physical space which is outside the pictorial one[49]. To avoid confusion, I will use "internal"/ "external viewer".

The internal viewer is a fundamental characteristic of primitive and pre-Renaissance art, Uspensky insists. A common representation of a garden in Egyptian art, for instance, creates the impression that the painter has placed himself mentally within the picture, "depicting the world as if it were *surrounding* him".[50] The mental movement from an outer to an inner position leads to a change of perception, which is the only way to make sense of "reverse perspectival" images. Uspensky wants to make us believe that, all of a sudden, "distortions" disappear and things come into place. Again we are faced with the already familiar concern of explaining "distortions" by denying them. There is nothing "distorted" about a

[49] Boris Uspensky, *The Semiotics of the Russian Icon*, (ed.) S. Rudy, (Lisse, 1976, first in Russian in 1971), p. 33.

[50] Ibid., p. 36.

"reverse perspectival" image once it is viewed from the correct position – this is what Russian theory on "reverse perspective" maintains. That this claim can in no way square with the other – of the representation of various aspects of an object which cannot be seen at the same time from a single point, no matter inner or outer – seems to go unnoticed.

Due to the importance assigned to the inner viewing position, Uspensky proceeds to bring forth evidence that indeed such a position is implied in ancient art and particularly in reverse perspective. His arguments can be grouped under three headings. The first has to do with the forms of mirror representation. The second condition deals with the source of light and the third with the size of the objects in relation to their distance from the viewer. I find all three conditions highly questionable.

Uspensky believes that in reverse perspective the image is not given as it would be seen by the viewer, but as it would appear reflected in a mirror. In this way, "the artist appears to be on the other side of the picture"[51] and not on the viewer's side. Uspensky maintains that icon-painting terminology bears witness to this phenomenon by considering as left-side what appears right to us, from our position before the picture and vice versa.[52] One of the author's examples to illustrate his statement is the medieval debate on St. Peter's and St. Paul's position in respect to Christ. The problem was whether the Apostles' places were to be determined considering the viewer or the image of Christ himself. According to Uspensky, here we have "two opposing artistic systems (the external and internal with respect to the representation)"[53]. There yet remains to be proved that the terminology Uspensky refers to implied an internal viewing position. Denoting "left" and "right" in relation to the represented object does not necessarily mean that the beholder has placed himself/herself inside the pictorial space.

In medieval art a common artistic phenomenon is the inner source of light which fades into shadow towards the front plane. Uspensky regards that inner light as a further indication of the respective positions of the beholder and viewer. He refers to an essay by Andrey Grabar on 'Plotin et les origins de l'esthétique médiévale' (1945).[54] Grabar connects the inner position of the medieval painter with Neoplatonic philosophy, especially Plotinus' belief that visual perception is built up not in the viewer's soul, but where the object is situated. In other words, the light metaphysics of Neoplatonism explains the source of light as an indication of a spiritual presence. It seems a wholly different matter to me, to jump from this idea to the conclusion that the pictorial image is shown as seen by the deity *and* as seen according to linear perspective. Along Neoplatonist lines, the shadow could

51 Ibid., p. 37.

52 Ibid., p. 36.

53 Ibid., p. 40.

54 See André Grabar, 'Plotin et les origins de l'esthétique médiévale', *Cahiers archéologiques: Fin de l'Antiquité et Moyen Age*, 1 (1945): 15–34 , reprint André Grabar, *Les origines de l'esthétique médiévale*, (Paris, 1992).

very well be interpreted as a veil cast over reality beyond which shines the light of truth, beauty and goodness.

Uspensky's third condition is the observation that the most obvious effect of reverse perspective, in stark contrast to linear perspective, is that the size of the objects decreases towards the front plane. This, according to the author, is a logical consequence of the inner position which determines the optical values of the image. There is no doubt that in reverse perspective there is a tendency of the lines to diverge in the distance. However, as I have mentioned several times already, this is not a sufficient condition to claim that the icon reverses the laws of linear perspective. Even though lines diverge in the first instance and converge in the second, they do so in a *different* manner. The condition would be valid only if the principle of convergence was the same, which would presuppose a single vanishing point in both cases, which we saw to be untrue.

Uspensky's discussion of the "internal viewer" of reverse perspective comes out of a wide background of intellectual ideas. Most obvious, perhaps, is the influence of Alois Riegl (1858–1905) in terms of terminology. Uspensky's "viewer" and "beholder" naturally brings to mind Riegl's concepts of "internal" and "external coherence" from his last work *The Group Portraiture of Holland* (1902). It deserves more attention to determine the kind of affinity that exists between Riegl's notion of the image, in which external coherence predominates, achieved, as it is, by "a rapport with the viewer"[55] and the "internal viewer" of reverse perspective, who is drawn into the pictorial space by the very nature of the spatial construction.

Uspensky mentions another idea from Riegl, namely the Austrian scholar's conception that in late Antique art the orientation was towards the object rather than towards the subject.[56] This notion is part of Riegl's interpretation of the development of art from a "haptic", objective view of things to an "optic", subjective one which owes a great deal to Hildebrand's "near" and "distant" view and is ultimately grounded in Hegel's aesthetics.[57] While the impact of these ideas on the Russian theory of reverse perspective deserves further study, for our purposes we will content ourselves with the framework set at the beginning of this chapter. Uspensky's theory is based on the ill-founded assumption that reverse perspective can be explained as actually reversing the laws of linear perspective. The whole notion of inner and outer viewing positions takes for granted that the basic concepts of linear perspective could function without modification within the pictorial space created by reverse perspective. Since Uspensky's position is based on concepts such as the vanishing point and the picture plane, it presupposes the construction of the visual pyramid of linear perspective. It is a further proof that

---

[55] Alois Riegl, *The Group Portraiture of Holland*, Los Angeles, CA, 1999), p. 220.

[56] Alois Riegl, *Late Roman Art Industry*, (Rome, 1985, first in German in 1901).

[57] See Michael Podro, *The Critical Historians of Art*, (New Haven and London, 1982), pp. 83–4; Margaret Iversen, *Alois Riegl: Art History and Theory*, (Cambridge, MA, and London, 1993), especially pp. 9–11, 44–6.

Gombrich was justified in saying that the "window" figure has won the world.[58] The application of these categories has been mechanical and I do not think it could prove justifiable on closer inspection.

An Alternative View: Karl Doehlemann

The line of thought on reverse perspective, as described so far, has been widely accepted. A recent book by Oleg Tarasov is no exception to the rule in that the author follows the guidelines, set by the trio Florensky – Zhegin – Uspensky.[59] It seems that Panofsky's Note 30 to *Perspective as a Symbolic Form*, an essay well-known in Russia as almost anywhere, has remained unnoticed. The note read:

> "The opinion of Oskar Wulff in *Die umgekehrten Perspektive und Niedersicht* [...] must be rejected on principle: namely, that 'reverse perspective' is a true inversion of normal perspective, in that the image is referred to a point of view of a beholder standing inside the picture instead of outside it."[60]

On this point, as on many others, Panofsky is not fully explicit. He does mention, though, that he sides with ideas expressed by Karl Doehlemann in his "Zur Frage der sogenannten 'umgekehrten Perspektive' in the *Repertorium fur Kunstwissenschaft*" 33 (1910). The Wulff–Doehlemann debate is of importance for us as it ultimately comes down to the very definition and legitimacy of "reverse perspective" as a concept. Moreover, to my knowledge, Doehlemann offers the only alternative view to Wulff's idea which has informed all major theories on reverse perspective.

Doehlemann opposes Wulff's notion of reverse perspective on the ground that there is a fundamental lack of autonomous abstraction of space with this perspectival system.[61] The compositional principle, as Wulff describes it, was unattainable since it requires a degree of geometrical-cum-optical abstraction achieved only with linear perspective. Byzantine art, according to Doehlemann, reveals no spatial coherence of its own. In that sense, it would be implausible to accept the theory of the inner view point, as it would imply the imposition of a spatial order which the "wholly naive" art form we are dealing with would

58 Ernst Gombrich, 'Western Art and the Perception of Space', *Storia dell'arte*, 62 (1988): 8.

59 Oleg Tarasov, *Icon and Devotion: Sacred Places in Imperial Russia*, (London, 2002, first in Russian in 1995), p. 12.

60 Erwin Panofsky, *Perspective as a Symbolic Form*, (New York, 1997), p. 114.

61 Karl Doehlemann, 'Zur Frage der sog. "umgekehrten Perspektive"', *Repertorium fur Kunstwissenschaft*, (Berlin, 1910), p. 86. I am grateful to Theodor Christchev of the Latin-English Medieval Dictionary, Oxford University Press for helping me with the text in German.

be unable to achieve. Recognizing that "reverse perspective" is a modern term that cannot adequately describe what "happens" in the pictorial space of the icon Doehlemann offers his own explanation for the phenomenon. He sees reverse perspective as an outcome of the hierarchical conception of relative size, where the size of the figures is seen in direct relation to the significance of each figure within the narrative. He suggests that all of Wulff's examples could be interpreted within that context.

Doehlemann's questioning of the term and concept of "reverse perspective" is fully justified, but both his idea that Byzantine art is "naive" and his explanation of its spatial organization are unsatisfactory. His simple and unitary explanation cannot deal with the complex optical and geometrical phenomena cited by Zhegin as aspects of "reverse perspective". Doehlemann's suggestion – it is not really a worked out theory – could well attempt to explain the "unrealistic" relationship among the objects in terms of size but not the perspectival treatment of each individual object. It comes down to the fact that objects of greater importance are very often given in an enlarged format in comparison to objects of lesser importance. Doehlemann refers to size as the indicator of status in the portrayal of royal personages. The hierarchical conception of the size of the depicted figures (i.e., the practice of showing the more important figures as larger in size than the less important ones), however, in no way explains the specific form of the Bible, for example, or the way in which space and form are actually indicated through various devices.

The worth of Doehlemann's essay lies, thus, not so much in his attempt at an explanation of the perspective in the icon, as in his challenge of Wulff's widely accepted view on the very notion of reverse perspective. To repeat again the thesis of the present text – the whole idea of defining reverse perspective through linear perspective is considered to be historically and philosophically unjustified, as well as optically and geometrically incorrect. What emerges as the most essential characteristic of space in icons is that it is *unsystematic* in optical terms and in that respect it is very different from the systematised pictorial space of the Renaissance.

Conclusion

The group of mainly Russian writers (too little-known in the West) forcefully press the claims of reverse perspective as truer to the processes of seeing than the standard form of perspective derived from the Renaissance innovations of Brunelleschi and Alberti. Their advocacy originated at a particular point in the development of early modernism, when traditional ideas of geometrical and physical space were breaking down both in science and art[62]. The "reverse perspective" of "ancient" Russian art was hailed as a prophetic kind of non-Euclidean geometry. It is evident that this argument suited the Russian aspirations of re-validating ancient Russian art as a counter to the Western naturalistic tradition. While reacting against Renaissance premises of art, much of the early twentieth-century theory of Eastern Orthodox art uses without modification concepts and categories deriving from Renaissance theory and practice. Thus, it fundamentally remains within the realm it purports to attack. The very notion of "reverse perspective" – as inverting the laws of linear perspective – is a telling example in that respect.

The major shortcomings of the theories on "reverse perspective" discussed here seem to be due to the adopted approach of describing the pictorial space in the icon in terms of linear perspective. Florensky implies that the value of Eastern Orthodox art lies largely in its adherence to the workings of human vision. Zhegin constructs his theory on the basis that each object in "reverse perspective" has its own vanishing point – a statement that is open to question. B. Uspensky follows O.Wulff's hypothesis of the inner – outer view, which is still built on what I believe to be a wrong premise.

Much more research needs to be undertaken on the problem of "reverse perspective" in order to develop new categories of analysis. As Sendler has pointed out, the existing literature on "reverse perspective" confirms that "research has not yet reached its goal"[63]. At this initial stage, the full implications of the need for a different approach to the problem should be realized. At a certain level, this may be seen as posing a danger to some of the basic assumptions of art history. It is widely believed that art history as a discipline grows out of the theoretical foundations for the study of Antique and Renaissance works of art. The major tool in that context becomes the understanding of linear perspective. As James Elkins maintains, art history is "dependent on two founding moments" – the Renaissance, which "remains the discipline's paradigmatic moment", and linear perspective as "the exemplary achievement of the Renaissance".[64] In a similar vein, Stephen Melville states that

> "the Renaissance achievement of rational perspective becomes the condition for the possibility of the art historical discipline, and we are compelled to its terms whenever we look to establish another world view that would not, for example, privilege the Renaissance, because we can neither 'look' nor imagine a 'world view' without reinstalling at the heart of our project terms only the Renaissance can expound for us".[65]

---

62 See Henderson, *The Fourth Dimension and Non-Euclidean Geometry in Modern Art*; Arthur Miller, *Einstein, Picasso: Space, Time and the Beauty That Causes Havoc* (New York, 2001).

63 Sendler, *The Icon: Image of the Invisible*, p. 148.

64 James Elkins, *The Poetics of Perspective*, (Ithaca and London, 1994), p. 189.

65 Stephen Melville, 'The Temptation of New Perspectives', *October*, 52 (1990): 11.

The idea that the concepts forged for privileging the Renaissance could just as well function for the explanation of categorically different art forms is difficult to accept. The case of "reverse perspective", I think, proves this assumption to be misplaced. Hegel's urge that we approach phenomena from the standpoint of their own historically relevant categories should be observed more consistently. The present book will suggest a possible approach to the problem of "reverse perspective" in terms which were valid at the time of the production of the images.

Chapter 3
Registering Presence in the Icon

When we look at images constructed in "reverse perspective", what inevitably strikes us at first glance is the lack of optical illusionism. So, what is the function of these images if it is obviously not a naturalistic representation of reality? The view proposed here is that "reverse perspective" can be regarded as an element of form which has been sanctified by tradition and, in this sense, as participating in the process of registering presence. More concretely, there is a *real*, *partial* presence of the prototype in the image which is witnessed, in visual terms, by the adherence to *canonical form.*

The Eastern Orthodox position on images is based on the belief that there is a sort of presence of the prototype in the image. In particular, there is an ontological identity, in a certain sense of the term, between Christ and the image of Christ.[1] For the modern reader, used to regarding images as aesthetic objects, such claims do not immediately make sense. A voluminous literature has grown round the attempt to explain and conceptualize the problem of the image as a carrier of presence. Recent publications on the subject testify to the importance of this issue and the possibility for further, illuminating interpretations, as well as the need to clear some of the misconceptions.

The term "real presence" is my choice of referring to the problem here by drawing an analogy with the dogma of the Real Presence in the Eucharist. The analogy became implicit during the Iconoclast debate. The Iconoclasts claimed that the only image of Christ was the Eucharistic bread and wine, thus assuming that the image had to be *homoousios* ("identical with", "essentially the same") with the prototype. The icon-defenders kept insisting that the Eucharist was not an image, but Truth itself.[2] It seems that they kept a closer connection with Platonic and Neoplatonic aesthetics, where the artistic image is only a reflection of the prototype and thus not fully and absolutely identical with it. In any case, it is apparent that the understanding of the nature of "*eikon*", its very definition, is different in either case. Jaroslav Pelikan points to the possibility of analyzing the "implications [of the nature of the Eucharistic presence] for the definition of 'image' and for the use of images".[3] The author puts the question in the following terms: "Was Eucharistic

---

[1] Later in this chapter we will come back to the question of why it is the image of Christ in particular which is of such crucial importance.

[2] Jacques-Paul Migne (ed.), *Patrologiae cursus completus. Series Graeca* (Paris, 1857-66), 99, 340B.

[3] Jaroslav Pelikan, *The Christian Tradition: A History of the Development of Doctrine*, vol. 2: *The Spirit of Eastern Christendom* (Chicago, 1971–1989), p. 94.

presence to be extended to a general principle about sacramental mediation of divine power through material objects, or was it an exclusive principle that precluded any such extension to other means of grace, such as images?"[4] The analogy is based on the idea that in both cases – in the Eucharist and in icons – there is, simply speaking, something present and something absent at the same time. There is a fundamental difference, however, which will emerge during the discussion. To use Aristotle's terminology, while the Real Presence in the Eucharist is *essential* (after the consecration, Christ is present in the elements of the Eucharist in his essence), in the icon the presence is *accidental* (the prototype is present in the image in its accident, but not in his essence). Or, in the Byzantines' own terminology, the image is similar to or *homoios* as the prototype, where *homoios* means "the same in quality, but not according to the essence/*ousia*"[5]. The identity between image and prototype can be understood in terms of *hypostasis* and not in terms of *ousia.*[6] I will use "real presence" with small letters in reference to the prototype-image relationship to distinguish it from discussions on Eucharistic presence where it is usually with capital letters.

The pictorial expression of the notion of presence is the adherence to canonical form, i.e., the accepted pictorial form, which has made its first appearance in the prototype. The form of the prototype – we will not be discussing questions as why this particular form came to be acknowledged as prototypical[7] – after a time becomes almost immediately recognizable by the believers through a multitude of copies. In a way, the act of recognition is a testimony of the validity of presence and it constitutes the recognition of the accepted pictorial form. As Charles Barber claims, the Byzantines provided a "formalist account of the icon",[8] since it appears that what prototype and image had in common was exactly form.[9] Naturally, formalism in the Byzantine context had a meaning which is very different from

---

4 Ibid.

5 Viktor Bychkov, *Vizantiiskaia estetika* (Byzantine Aesthetics) (Moscow, 1977), p. 29; the translation is mine. On the distinction between *homos* and *homoios*, see the Glossary.

6 It is to be noted that till the fourth century there was no difference between *hypostasis* and *ousia*, while in Latin both terms have been translated as *substantia* (*op. cit.*, p. 29). *Ousia* (nature, substance) refers to that which is held in common, while *hypostasis* refers to "a reality whose specificity is signified prior to that of common nature" (Alain Besançon, *The Forbidden Image: An Intellectual History of Iconoclasm* (Chicago and London, 2000, first in French in 1994), p. 116). Thus, the Trinity shares a common *ousia*, but three different *hypostases* (see 'Letter 38' from St.Basil to his Brother Gregory of Nyssa').

7 Whitney Davis's thesis that the Egyptian canon was a matter of social choice made out of a variety of possible alternatives could make sense here, too (Whitney Davis, *The Canonical Tradition in Ancient Egyptian Art* (Cambridge, 1989).

8 Charles Barber, *Figure and Likeness: On the Limits of Representation in Byzantine Iconoclasm* (Princeton and Oxford, 2002), p. 110.

9 Ibid., p. 117.

that of modern aesthetic formalism. Barber is aware of this distinction and says that form was not "an independent aesthetic attribute" for the Byzantine theologians, but rather "a cause for the icon".[10] Form in this sense (the Greek *character*, but also one of the connotations of *morphe*; see Glossary) is synonymous with "accident", which brings us back to Aristotle's terminology. In *De Anima*, II.12.424a Aristotle says that "species" or "sensible forms", emanated by objects imprint themselves on the wax-like receptacles of our senses like a signet ring. The classical Greek background of that idea is witnessed by the fact that the ancient Greek term *character* means both seal and the impression left by a seal. At the heart of this idea, among other things, is a concept of fundamental significance in the theology of the icon as well as in aesthetics, which is that of pure form. Form can be detached from substance and transferred onto another substance and thus it becomes the bearer of identity (see the term "form" in the Glossary).

The questions that will be discussed in the following text – the cult of images in the Eastern Orthodox world and in the West, the relationship between icon and relic, between prototype and image, between word and image – are meant to show the origins and some of the main components of the view, concerned with the problem of presence in the image. Further, this view does not pretend to offer a thoroughly logical and consistent position. On the contrary, it actually seems that the Byzantines deepened the paradox at the heart of the image, inherited from pagan Neoplatonism and understood very much in the sense of that Kierkegaard later imparted to the term. According to Kierkegaard, Christianity is deeply and inherently paradoxical and it should be understood and accepted as such – the co-existence of contradictory aspects, impossible to harmonize according to the strict laws of reason.[11] This view is close in spirit to a major trend, running throughout Eastern Orthodox thought to the present, which stresses the insufficiency of logic to resolve the antinomies at the heart of Christianity.[12] The problems under our attention, belonging to the field of the theology of the image, will be seen to reflect the antinomy "transcendent – immanent" and so the absolute paradox in the nature of Christ. The Eastern Orthodox tradition, dealing with the problem of images, treats the icon in this manner – as a paradoxical unity of the transcendental and the immanent.

---

[10] Ibid., p. 119. On this, see also Golbert Dagron, 'Holy Images and Likeness', *Dumbarton Oaks Papers*, 45 (1991): 23–33.

[11] See Kierkeggard's *Fear and Trembling* (1843), *Repetition* (1843), *Philosophical Fragments* (1844), *Concluding Unscientific Postscript* (1846) and *Practice in Christianity* (1850).

[12] The roots are to be sought in the crisis of Hellenistic rationalistic thought, which provided the background to patristic philosophy. This theme is prominent particularly with the Cappadocians, Pseudo-Dionysius, Maximus the Confessor, etc. In certain respects a parallel could be drawn with the crisis of modernity, following the Enlightenment, as we will see with Florensky.

Philosophically, the Byzantine understanding of the icon, as worked out during the First (726–787) and Second Iconoclastic period (814–843), was based on pagan Neoplatonic theory of the image. What is less known is that it re-appears in interesting ways in Russian religious philosophy. Florensky's writings from the 1920s consider the image largely under the guise of a romantic definition of the symbol. Thus, Florensky belongs both to a rich Eastern Orthodox tradition and also to an intellectual movement at the beginning of the twentieth century which revived romantic theories in the light of a modern concern with the crisis of modernity.

The critical literature, devoted to the theology of the icon is huge.[13] The Iconoclast Controversy in Byzantium, in particular, has drawn a lot of attention in recent years.[14] This has been especially true on part of Western scholars, who have tended to see Byzantine Iconoclasm as an ancestor of developments in Western Europe (the Reformation, ideas of the French and Bolshevik Revolutions, etc.).

The Cult of Images and Eastern Orthodox Identity

The cult of images has been a persistent characteristic of Eastern Orthodox civilization. It has endured through the centuries and has been strengthened into a dogma of Eastern Orthodox faith. In 726 dozens of pious women defending the icon of Christ from the Chalke Gate in Constantinople from destruction were put to death at the order of Leo III, the first Iconoclast Emperor.[15] The date marks the outbreak of Iconoclasm which lasted from 726, with an intermission,[16] to 843, but also the deeply felt popular attachment to icons. History and legend tell us of numerous stories of individuals who would not renounce their worship of icons, even at the expense of their lives.[17] The end of Iconoclasm asserted once and for all

[13] A useful anthology of selected texts by Byzantine and modern Russian authors can be found in Kalistos Ware, 'The Theology of the Icon: A Short Anthology', *Eastern Churches Review* (1976-7): 3–9.

[14] For a select bibliography see Andrew Louth, 'Iconoclasm' in Adrian Hastings, Alistair Mason, and Hugh Pyper (eds.), *The Oxford Companion to Christian Thought* (Oxford, New York, 2000), p. 318. An important recent work is Barber's *Figure and Likeness: On the Limits of Representation in Byzantine Iconoclasm.*

[15] The image of Christ of the Chalke, according to the *Vita Constantini*, had taken the place of portrait of the Emperor Constantine. The historicity of the famous Chalke Gate episode has been called into question by recent scholars, as, for example, by Marie-France Auzépy in 'La destruction de l'icône du Christ de la Chalcé par Leon III: propagande ou réalité?', *Byzantion*, 60 (1990): 445–92.

[16] Imperial policy was reversed in 787–814 under the Empress Irene.

[17] An example is 'The Life of St. Stephen the Younger' from around 800 (See 'The Life of St. Stephen the Younger' in A-M Talbot (ed.), *Byzantine Defenders of Images: Eight Saints' Lives in English Translation* (Washington, DC, 1998, pp. 9–13). While this does not come across in the *Vita*, Stephen the Younger was actually executed for his implication in a

the power of the visual in Eastern Orthodox religion. Since 843 and to the present the faithful celebrate one of most important feasts of the Orthodox Church, "the Triumph of Ordodoxy" (on the first Sunday of Lent), in honour of the victory over Iconoclasm. The importance of the victory of the Iconophile party, however, goes beyond its immediate theological implications. It signals nothing less that the immense role that the icon was going to play in forging an Eastern Orthodox cultural identity. As Robin Cormack maintains, "from 843 onwards, the position was that to deny the icon was to deny the identity of the Orthodox believer".[18]

The cult of images, however, had been intensified to a high pitch long before the eighth century. According to Ernst Kitzinger this phenomenon had already reached great proportions in the era after Justinian and especially after the second half of the sixth century[19]. There is evidence of religious images that were subjected to a variety of elaborate devotional practices. The first literary record of *proskynesis* (prostration)[20] in front of images in churches dates from the first half of the sixth century in a letter written by Bishop Hypatius of Ephesus.[21] The *Life of St. Symeon the Younger*[22] (d. AD 592) brings to attention the use of candles before images[23] and in the acts of Nicaea II (787) "the offering of incense and

conspiracy against the Emperor (Patricia Karlin-Hayter, 'Iconoclasm' in Cyril Mango (ed.), *The Oxford History of Byzantium* (Oxford, 2002), p. 158). This allows us to see the deep intertwining between politics and religion in Byzantium.

[18] Robin Cormack, *Painting the Soul Icons, Death Masks and Shrouds* (London, 1997), p. 43.

[19] See the 'Cult of Images in the Age before Iconoclasm' in Ernst Kitzinger, *The Art of Byzantium and the Medieval West* (Bloomington, London, 1976). Also, Averil Cameron, 'The Language of Images: the Rise of Icons and Christian Representation' in Diana Wood (ed.), *The Church and the Arts* (Oxford and Cambridge, MA, 1992), pp. 1–43.

[20] The origin of *proskynesis* before images is to be sought in the prostration in front of the Emperor, which is contrary to Roman practice and derives from Oriental (particularly Persian) court ceremonial. Procopius comments on that practice and sees it as infringement of Roman liberties (Procopius, *The Secret History* (Harmondsworth, 1981), p. 192). Actually it was Diocletian (284–305), who was the first emperor to demand homage in the form of prostration. The attitude to the persona of the emperor that this implies has much earlier antecedents, however (See Averil Cameron, *The Late Roman World: 384–430 AD* (London, 1993), p. 42).

[21] Kitzinger, *The Art of Byzantium and the Medieval West*, p. 100.

[22] The *Life* was written probably in 809, i.e., at a period when the iconophile cause was victorious. It refers the reader back to a time when Iconoclasm was at its height. Marie-France Auzépy believes that the text has been constructed on the basis of extracts from earlier lives (for instance, by Andrew of Crete and Cyril of Scythopolis) and from the conciliar *Acta* (Nicaea II) (Marie Auzépy, *La vie d'Étienne le Jeune par Étienne le Diacre* (Aldershot, 1997).

[23] Kitzinger, *The Art of Byzantium and the Medieval West*, p. 102.

lights" to images as a "piously established ancient custom"[24] is mentioned. Prayers were being addressed to icons at least since the time of St. Augustine.[25] A text that cannot be precisely dated but roughly belongs to the late sixth and seventh century bears witness to the degree of intensity that the worship of images had reached at the time. It gives account of a group of workmen paying their devotion to the image of the Virgin – they "embraced it and kissed its hands and feet and continued to salute it a long time pressing it to their bosoms in great faith".[26] Maximus the Confessor (*c*.580–662) speaks of the kissing (*aspasmos*) of the icons of Christ and the Virgin on special occasions.[27]

The presence of images was ubiquitous not only in the private, but in the public sphere, as well. Icons would be placed on or near public gateways, as in the case of the already mentioned Christ of the Chalke, the main gateway to the imperial palace in Constantinople. *The Life of St. Symeon the Elder* tells of images of the saint put over doors of workshops. The use of religious images as *palladia* (from *palladium*, a public cult object recognized by all) especially in war time was wide-spread. This practice had its direct antecedent in the worship of images in the domestic sphere but it has also more ancient pagan roots. There are several accounts of the role played by the *mandylion*, the legendary image of Christ at Edessa during the Persian attack on the city in 544.[28] During the siege of Constantinople by the Avars in 626 icons were placed above the gates of the city and carried in solemn processions around the walls.

It seems that the surviving literary and historical evidence justifies the view that the veneration of images was adopted in the sixth century. It is during this period – from the sixth to the eighth centuries – that Byzantine society underwent deep cultural changes. The result was the disintegration of the Late Antique worldview and its gradual replacement by a new one which was typically medieval and Byzantine. As Averil Cameron says, religious images were "one element in the necessary construction of an alternative worldview".[29] Actually the author convincingly proposes in her article that images were "part of an urge to assert a new authority",[30] i.e., they filled in a vacuum left by the breaking down of the

[24] Norman Tanner (ed.), *Decrees of the Ecumenical Councils* (London and Washington, 1990), vol. 1, p. 136.

[25] Kitzinger, *The Art of Byzantium and the Medieval West*, p. 104.

[26] Ibid.

[27] See text 7B in Hans Belting, *Likeness and Presence* (Chicago and London, 1994), p. 505.

[28] The mandylion was allegedly sent to King Abgar by Christ himself. The first account of this legend is found in Evagrius (b.536) (See Scholasticus Evagrius, *The Ecclesiastical History of Evagrius: With the Scholia* (Amsterdam, London, 1898, rpt.1964).

[29] Cameron, 'The Language of Images: the Rise of Icons and Christian Representation', p. 34.

[30] Ibid., p. 4. See also Robin Cormack, 'Byzantine Aphrodisias. Changing the Symbolic Map of a City', *Proceedings of the Cambridge Philological Association*, 216/36

culture of Classical Antiquity. In this sense, it could be claimed that by the ninth century, Byzantine culture came into its own. Most importantly within the context of our discussion, this new cultural awareness was "expressed through and with icons".[31] Thus, the most visible manifestation of identity was that "henceforth, church, home, and street were all the domain of the icon".[32]

For centuries the icon was to be a permanent element in the make-up of Eastern Orthodox identity. This partly explains the striking persistence of devotional practices and attitudes towards sacred images. In Tolstoy's *War and Peace*, for instance, the icon of the Smolensk Mother of God is treated very much like a medieval *palladium.* It is carried along with the Russian army and solemnly paraded in procession before a battle. Soldiers and officers bow in front of the image and kiss it in a manner that recalls the medieval *proskynesis* and *aspasmos.* These are all aspects of a ritualistic behaviour which had come to be identified as typically Eastern Orthodox. We read of the old General Kutuzov, who "went to the icon, sank heavily on his knees and bowed to the ground" and then "put out his lips in a naïve, child-like way and kissed the icon, and again bowed down".[33] The Smolensk image, in a sense, becomes for the Russians in the nineteenth century what the *mandylion* had been for the Byzantines in the sixth. They are treated in similar ways and the connection between the two is exactly the sense of Eastern Orthodox identity embodied by the image.

The Western Position – a Critical Reappraisal

The idea that the cult of images is a permanent factor in the make-up of Eastern Orthodox identity has all too often led to the exaggerated contrast between Eastern Orthodox and Western attitudes to images. This tendency could be seen as one of the manifestations of the overall concern to highlight differences between Orthodox and Catholic theology. A closer inspection of attitudes, however, throughout the Middle Ages and at least up to the Reformation, reveals that the Orthodox East and the Latin West had fundamentally the same attitude to religious images, while divergences were frequently blown out of proportion at the time and later.

There were indeed historical precedents, both before and after the Schism in 1054,[34] that justify Stephen Runciman's opinion that "Eastern and Western Christendom had been drifting apart in theology, in liturgical usage and in

(1990): 26–41.

[31] Cormack, *Painting the Soul*, p. 151.

[32] Ibid., p. 166.

[33] Leo Tolstoy, *War and Peace* (London, 1968), p. 919.

[34] The formal schism between the Churches took place in 1054 when Cardinal Humbert and two other papal legates placed a bull of excommunication on the altar of Hagia Sophia in Constantinople.

ecclesiastical theory and practice".[35] The differences were felt to be so acute that one of Byzantium's last ministers, Lucas Notaras, reportedly said: "Better the Sultan's turban than the Cardinal's hat".[36] Without making such a comparison, the Russians apparently shared the sentiment behind it. The Metropolitan of Moscow commented in 1458 that: "Constantinople has fallen because it has deserted true Orthodox faith", referring to the Council of Ferrara-Florence of twenty years earlier, at which the Byzantine delegation had been pressed to sign a union of the Churches, mostly along the Catholic lines.[37]

Specifically in relation to the theology of the image, the notion of contrasting attitudes has had a long history in critical literature. According to Ernst Kitzinger, one of the foremost Byzantine art historians, Pope Gregory's two letters to the Bishop of Marseilles of *c.*600 are the "classical expressions of the Western attitude".[38] Pope Gregory's reaction towards religious images is a balanced one – while they should be tolerated and utilized as educational tools they should not be worshipped as sacred objects. The same attitude was found among Frankish theologians and was expressed in the famous *Libri Carolini*, an expanded form of the now lost list of objections to the Acts of Nicaea II, which Charlemagne sent to the Pope. In the West, Kitzinger sums up, icons were said to be appropriate for ornamental and educational purposes but were not to be shown liturgical honour, as in Byzantium. The distinction between a "Western"/ Frankish and "Eastern" view of icons is clearly put forward in the foreword to the catalogue of the 1987 London exhibition of icons.[39] Indeed, there has been a wide-spread tendency to identify the position of the West as a whole with the views expressed in the *Libri Carolini*. As a result, the Western view appears to be a simplified one, where the icon is no more than *biblia pauperum* in Pope Gregory's expression. Consequently, it lacks all the sophistication that a Christological argument gives to the debate in the Eastern Empire. Whenever the contribution of Rome comes to light it is interpreted in that context. Thus, Michel Quenot sees "a profound misunderstanding" between the conceptions of religious art in the East and in the West. The "serious discrepancy" the author attributes to the "divergent opinions about the essential element of the icon – its *theology of presence*".[40] In the same vein, Gerardus van der Leeuw

[35] Stephen Runciman, *The Fall of Constantinople 1453* (Cambridge, 1963), p. 7.

[36] Ibid., p. 21.

[37] In fact, the Eastern patriarchs refused to stand by the decision of their delegation.

[38] Kitzinger, *The Art of Byzantium and the Medieval West*, p. 138.

[39] Kallistos Ware, 'The Theology and Spirituality of the Icon' in Royal Academy of Arts, *From Byzantium to El Greco: Greek Frescos and Icons*, London, 27th March – 21st June, 1987 exhibition, p. 37.

[40] Michel Quenot, *The Icon: Window on the Kingdom* (London, 1992, first in French in 1987), p. 79.

maintains that "the West sees in images only instruction, edification", while the East – "mysteries, which affect salvation".[41]

More recently, Bronwen Neil has called these views in question, thus lending support to Hans Belting's well-known *Likeness and Presence* (1990), which draws attention to similar, even identical attitudes towards the sacred image in the West and Byzantium "before the era of art".[42]

Neil's article is concerned with 'The Western Reaction to the Council of Nicaea II'. Nicaea II, or the Second Ecumenical Council of 787, took place during the First Iconoclastic period and arguably it represented "the most complete teaching yet given on the icon".[43] Neil proposes that there is a fundamental difference between the positions of the Papacy and of the Frankish Church respectively on the question of icons and that the former is actually very supportive of the Byzantines' view.[44] The author goes further by maintaining that actually Pope Hadrian I (772–795) actively contributed to the decisions reached by Nicaea II (787) and that the papal documents presented at the Council express the official position of the Church of Rome. The negative response of the Franks, the author suggests, was the result of their misunderstanding of the basic Greek arguments[45] and was largely due to imprecise and often wrong translation of the Acts of the Council.[46] More particularly, the two Greek terms – *time* or "veneration" (offered to images) and *latreia* or "worship" (offered to God alone) were translated into Latin, in both cases, as "adorare". Thus, the fine distinction at the basis of the definition of icon worship was lost to the Franks, whose attitude remained consistently hostile. In fact, the Frankish Church formally condemned icon worship – at the Synod of Frankfurt (794) and at the one of Paris (825).

It seems that the importance of the *Libri Carolini* as the statement of the Western position on images has been exaggerated too much and that actually they had a "limited influence"[47] on the whole. The life of the Frankish Church itself was short-lived and certain of its views are representative exclusively of Frankish theology, while the West at a later period seems to have broken away from them.

---

41 Gerardus van der Leeuw, *Sacred and Profane Beauty: the Holy in Art* (New York, Chicago, 1963, first in Dutch in 1932), p. 175.

42 The sub-title to Belting's book is *A History of the Image in the Era before Art.*

43 Robinson, *Images of Byzantium*, p. 8. On the Nicaea II and its position on images see Ambrosios Giakalis, *Images of the Divine: Theology of Icons at the Seventh Ecumenical Council* (Leiden, New York, 1994), especially pp. 14–20.

44 Bronwen Neil, 'The Western Reaction to the Council of Nicaea II', *Journal of Theological Studies*, New Series, 51 (2000).

45 Ibid., p. 533.

46 Ibid., p. 549. The imprecise translation of the Acts of the Council and the misunderstanding it caused in the West is noticed by various scholars. See, for example, Giakalis, *Images of the Divine: Theology of Icons at the Seventh Ecumenical Council*, p. 21.

47 Belting, *Likeness and Presence*, p. 298.

The Pope's reply to Charlemagne (793?), however, is a detailed defence of the Iconophile position. It is thus a reassertion of the Pope's views, as outlined in the two letters, included in the Acts of the Council (787). Neil makes a convincing case for her thesis that there was a Roman contribution to Nicaea II and that Hadrian's apology for icon worship makes use not solely of the pedagogical argument, but also of the Christological and the one stressing the role of tradition. The latter two arguments are usually considered as a purely Byzantine ground.

The two views on images in the West – the first associated with Rome and having a lot in common with the Eastern Orthodox one and the second with the Frankish Church, which was antagonistic to both Rome and Constantinople – are significant in a wider historical context as they forecast future developments which led to the Reformation in the West. One of the great iconoclasts of the Reformation, Calvin, explicitly rejected Nicaea II and based himself on the *Libri Carolini*. However, in the period under discussion it is the Roman position that eventually won the day.

As Belting shows, by the early seventh century the veneration of images was a common practice in Rome as well and the forms that the cult of images took were frequently similar, sometimes identical, to those in Byzantium.[48] The annual procession of the icon of Christ to S. Maria Maggiore in Rome was an established custom by the eighth century.[49] It is among the instances of such events that were widely spread at the time, as "many other images are carried in processions".[50] Throughout the Middle Ages, there are even instances of a very similar attitude in the Eastern Empire and in the West provoked by the same image. The icon of the Virgin *Nicopeia* (from the eleventh century) was an honoured *palladium* for the Byzantines and was regarded as their celestial commander in times of war and peril. When the Crusaders seized it, it was brought with great ceremony to Venice and became one in a line of images claiming to be St. Luke's portrait of the Virgin. It was treated as the true sovereign of the state and received the honours of such.[51]

The West frequently borrowed elements from Byzantine legends about icons. There were certain motifs that created an aura around an image and some of them were the same both in Byzantium and the West. Probably the most popular Western image, the Veronica, kept in St. Peter's in Rome, first made its appearance in the early thirteenth century. The story clearly derives from the Abgar legend, which is of Byzantine origin – Christ impressed his features on a cloth, offered to him by a woman, called Veronica. Dante in the *Vita Nuova* tells of the multitudes of pilgrims going to Rome "to see the blessed image that Jesus Christ left us as a copy of His most beautiful face (which my lady beholds in glory)".[52]

---

48 Ibid, p. 124.

49 See text 4B in Belting, *Likeness and Presence*, pp. 498–9.

50 See text 4F in Belting, *Likeness and Presence*, p. 500.

51 Belting, *Likeness and Presence*, p. 204.

52 Dante, *Vita Nuova* (Oxford and New York, 1998), p. 80.

The views of the Church of Rome seem to have continued at least till the Reformation. Leonardo described the reaction to the sacred image in Renaissance Italy, which carries immediate reminiscences to Eastern Orthodox devotional practice. It is interesting to notice the idea of real presence in the image:

> "Do we not see pictures representing divine beings constantly kept under coverlets of the greatest price? And whenever they are unveiled there is first great ecclesiastical solemnity with much hymn singing, and then at the moment of unveiling the great multitude of people who have gathered there throw themselves to the ground, worshipping and praying to the deity, who is represented in the picture, for the repairing of their lost health and for their eternal salvation, exactly as if the goddess were there as a living presence".[53]

As much as there are differences between Byzantine and Western attitudes towards images, these do not seem to be indicative of any major ideological distinctions. For example, the Catholics have rarely used the posture of prostration before images to express their respect. The only exception to this is the Good Friday ceremony of the adoration of the cross, but it is telling, as it shows that the Catholic Church has nothing against the practice of *proskynesis* before images in principle. A difference that comes across most forcibly is probably a matter of a shift of emphasis or rather, the amount of attention that certain issues received in the West and in Byzantium respectively. The question is not that the Catholic Church thought any differently of the nature of the relationship between prototype and image. Rather, it did not problematize the issue as much as Byzantine theology did. As Alain Besançon points out, "in the West, there was never a debate about the divine image comparable in depth, extension, detail, and violence to that which had long occupied the East".[54] In this sense, we can speak of "a profound difference in intensity" alongside "a fundamental [theological] parallelism".[55] Images were just as much invested with the powers to represent a legal person, but as Belting puts it, this was "a custom, not a philosophical problem".[56] Western legends about miraculous images tend to focus more on the age of images that on their miraculous origin. They rarely dwell on proofs for the authentic reproduction of a prototype in the image.[57]

From what has been said so far, it seems legitimate to accept that the visual image plays a similar role both for the Eastern Orthodox and the Latins. A fundamentally different attitude can be noticed only after the Reformation on two

[53] Martin Kemp (ed.), *Leonardo on Painting* (New Haven and London, 1989), p. 20; from the *Codex Urbinas*, 2v – 3v (McM 17–18).

[54] Alain Besançon, *The Forbidden Image: An Intellectual History of Iconoclasm*, pp. 146–7.

[55] Ibid, p. 148.

[56] Belting, *Likeness and Presence*, p. 305.

[57] Ibid, p. 308.

levels. Firstly, though in varying degrees, the Protestant Churches promoted an iconoclast understanding of the image. Secondly, more or less at the same time, the image in its religious dimension was giving way under the pressure of the art product as an object of aesthetics.

Icon and Relic

Studies on the relic invariably emphasize the dimension of presence, i.e., the belief that the saint is somehow present in his relic, which is witnessed by the miraculous properties that relics are frequently believed to display. Therefore, an analogy with the image which, too, contains presence is almost unavoidable. What usually escapes attention is that there is a fundamental difference relating to the terms on which presence is predicated. What is important in the case of the relic is that it is of a material which has been in contact with the saint.[58] This idea might be valid with some images, as well. The most obvious example would be the so-called *acheiropoietoi*[59] or images not made by human hands. The most famous of those is probably the mandylion or Christ's image which Christ himself impressed on a cloth which he then sent to King Abgar of Edessa.[60] However, Gary Vikan's statement that an icon is alike to the relic in so far as it is an "image not created by human hand but through a miracle"[61] seems unconvincing. The majority of sacred icons cannot and do not lay such a claim which does not prevent the belief that they do disclose presence. Apparently, in the case of the icon we need another, distinct notion that would guarantee the presence of the prototype. This notion is form and it is specific to the image, in contrast to the relic, where it plays no role. As David Freedberg says, "it is precisely the figural aspect of images that distinguishes images from relics *tout court*".[62]

However, as both relic and image have been seen as containers of presence they frequently provoked a similar attitude. This has led authors to suggest that in Eastern Orthodox Christianity the icon gradually assumed the position occupied by the relic. This view has been proposed by the Russian Byzantinist Andrey Grabar

58 Gary Vikan, 'Relics and Icons', in Myrtali Acheimaston-Potamianon (ed.), *Holy Image, Holy Space*, exhib.cat. (Athens and Baltimore, 1988), pp. 45–7.

59 In Latin "non manufactum". The same concept persists in post-Byzantine times, as the Russian word "nerukotvornii" shows. Ernst von Dobschutz has shown that *acheiropoietoi* have their antecedents in *diipeteis* or objects that the Greeks believed were cast down by Zeus and interprets them as "a case of ancient beliefs being transferred into Christian concepts". (Dobschutz, E. von, *Christusbilder* (Leipzig, 1899), pp. 256–66).

60 The image was brought to Constantinople in 944 from Edessa, Northern Syria and installed in the palace chapel. After 1204 we lose track of it.

61 *op. cit.*

62 David Freedberg, *The Power of Images* (Chicago and London, 1989), p. 97.

in his classical study on the subject.[63] Karel Innemee suggests in the same vein that the production and veneration of icons arises "at least partially from the cult of the relic".[64] The change of emphasis in Byzantium began, in Ernst Kitzinger's opinion, in the late sixth and seventh centuries,[65] when the icon started to replace the relic as a principal object of devotion. In the preceding period we encounter forms of the worship of relics that are strikingly similar to later devotional practices before images. Gregory of Nyssa (*c*.335 – *c*.394) describes the following scene: "Those who behold them [the relics] embrace, as it were, the living body itself in its full flower, they bring eye, mouth, ear, all their senses into play, and then shedding tears of reverence and passion, they address to the martyr their prayer of intercession as though he were hale and present".[66] As we saw, this is ritualistic behaviour that was to become closely associated with image veneration.

The near overlapping of icon and relic can be witnessed in an extreme form in the *eulogia*, amulets brought back by pilgrims to the Holy Land. These amulets are very often images on the outside that contain relics in an inner compartment. And conversely, there are cases when the opposite happens – relics are actually treated as images. Peter Brown draws attention to the instance of the corpse of Daniel Stylites (d. 493) in Constantinople that had been fixed to an upright panel and venerated "like an icon".[67] Actually, according to Belting, "no significant distinctions were made between image and relic, since cult objects were by nature relics, even if they were not parts of bodies".[68] Later in his book Belting claims, that "in medieval imagination images and relics were never two distinct realities".[69] Although, in the long run, the icon and the relic became two separate phenomena[70] the link between them provides a useful insight into the nature of both. More so, if we accept that "in the realm of personal devotion the connection [...] has never really disappeared".[71] This connection was sustained by the notion of presence. At the same time, as was mentioned already, presence was disclosed by different means. This is why, however close the connection between icon and relic, the cult

63 See André Grabar, *Martyrium: reserches sur le culte des reliques et l'art chrétien antique* (Paris, 1946).

64 Karel Innemée, 'Some Notes on Icons and Relics', in Christopher Moss and Katherine Kiefer (eds), *Byzantine East and Latin West* (Princeton, 1995), p. 520.

65 Kitzinger, *The Art of Byzantium and the Medieval West*, p. 121.

66 I am quoting from ibid., p. 122.

67 Peter Brown, *Society and the Holy* (Chicago, 1982), p. 251, pp. 266–7, pp. 275–84.

68 Belting, *Likeness and Presence*, p. 195.

69 *op. cit.*, p. 300.

70 According to Kitzinger (See Kitzinger, *The Art of Byzantium and the Medieval West*, p. 122) the emancipation of the icon from the relic was almost complete by the end of the seventh century.

71 Innemee, 'Some Notes on Icons and Relics', p. 521.

of images was never "entirely dependent on"[72] on the cult of relics. The sacred image had developed its own conception of presence. The medieval icon borrowed this conception from the Late Antique image, while it further emphasized and elaborated on the idea of identity as shared form. This view ultimately committed authors to the notion of partial presence – i.e., the identity between prototype and image is only in certain but not all aspects.

The Real Presence in the Image

As St. John of Damascus puts it "when we venerate images, it is not veneration offered to matter, but to those portrayed through matter in the images. Any honour given to an icon is transferred to its prototype".[73] It could be maintained that the whole of the Iconoclast controversy in Byzantium rotated around the question of the relationship between prototype and image. The image of Christ is the main paradigm which defines the parameters of the whole debate on Christian imagery.

The spectrum of different views fluctuates from one end, where the image is seen as fully identical with the prototype, to the other where the image is only a sign for another higher reality and the ontological connection between the two is completely severed. The position of the Eastern Orthodox Church seems to be somewhere in between these two extremes and it will be the topic of this sub-chapter. The concept of the real presence of the prototype in the image will be discussed in order to illuminate the nature of the problem under our attention.[74]

As St. John's statement above shows, the aim of the image in Eastern Orthodoxy is to serve as a pathway to the prototype. Moreover, this becomes possible only because the prototype is, in some way, present in the image. This belief is well illustrated by St. Symeon when he says: "When we see the Invisible through the visible picture we honour Him as if He were present".[75] We will attempt to see *how* exactly this happens and what constitutes the nature of the presence of the prototype in the image.

The rationale of this view of the image was first worked out by pagan Platonism and Neoplatonism. It was in Late Antiquity, too, that we should look for the notion of portrait-type, i.e., a certain accepted and recognizable form that acts as a dominator of presence.

---

[72] Kitzinger, *The Art of Byzantium and the Medieval West*, p. 125.

[73] John of Damascus, *On the Divine Images* (Crestwood, New York, 1980), p. 89.

[74] Probably the best account of the relationship between prototype and image in Antiquity and in Byzantium is in Moshe Barasch, *Icon: Studies in the History of an Idea* (New York and London, 1992).

[75] Kitzinger, *The Art of Byzantium and the Medieval West*, p. 153. The quote is from a sermon ascribed to the saint.

Classical Antique Sources

According to Belting, the medieval icon directly derives from three traditions of classical panel painting – the divine image, the imperial image and the portrait of the dead.[76] We can see how certain genre characteristics of the icon, both artistic and ideological, are intimately connected to its origins. A point of interest is whether, if at all, presence was presumed on the basis of actual physical resemblance and if/when not what notion gave grounds for the assumed reality of presence.

Origins of the Icon in Late Antiquity – Presence Based on Type

The idea that a person can inhabit his image has pagan origins. A story by Pausanias (AD 143–176) tells of how the statue of the Thasian athlete Theagenes fell on a man, who having hated Theagenes during his lifetime had come at night to the former athlete's statue "to flog the bronze as if he were beating Theagenes himself".[77] As the person was killed as a result, his sons proceeded to prosecute the statue for murder and "the Thasians took the opinion of Drakon, who rules in Athenian murder laws that even inanimate objects which fell on a man and killed him must be taken outside the boundaries, and they drowned Theagenes' statue in sea".[78] It was, then, not just a matter of superstition shared by sections of society, but an official practice to take legal action against a person's statue if, for some reason, the man himself could not be found.

With the so-called Fayum portraits of the dead of the Roman period[79] the mummy and by implication the portrait could well be regarded "as the immortal surrogate of the deceased".[80] A mummy label preserves a note from Didymos' father, who refers to the mummy as "my son" as if he were still alive.[81] Physical resemblance seems to have been an important element of this belief. A significant number of the more than a thousand portraits that have been recovered disclose a strikingly naturalistic style that comes in a direct line from illusionistic art of the Hellenistic era. Recently scholars have tended increasingly towards the view, expressed already by the first pioneers in uncovering the Fayum paintings. At the turn of the century the French Egyptologist Albert-Jean Gayer tells of the comparisons he would make between a mummy's face and the portrait inserted in the mummy. His conclusion was that "despite the desiccation of the flesh, it

---

76 Belting, *Likeness and Presence*, p. 26 and p. 78.

77 Pausanias, *Guide to Greece*, vol.2, book VI (Harmondsworth, 1971), p. 316. A similar story of a statue falling to take revenge can be found in Aristotle's *Poetics*, 9, 12.

78 Ibid.

79 After the Fayum valley in Egypt, where the greatest number of such artefacts were found.

80 Euphrosyne Doxiadis, *The Mysterious Fayum Portraits: Faces from Ancient Egypt* (London, 1995), p. 39.

81 Ibid., p. 85.

was easy to recognize the features". This was particularly evident in the hairstyle – the portrait would have "exactly the same coiffure as worn by the deceased, not the slightest curl was misplaced".[82] Gayer seems to be implying that the portrait was taken after death, which could not have always been the case. Actually Flinders Petrie's much debated theory, based on portraits he found in the acropolis of Hawara in the Fayum in 1888, was that most of the portraits were painted while the subjects were still alive.[83] This seems to be confirmed by the fact that the majority of the portraits show youthful people. Cormack believes that this phenomenon reveals "an obvious warning against the interpretation of realism" so long as we do not mean to imply that "Romans in Italy in the Republic lived to an older age and status than the ruling classes of Egypt".[84] However, if we accept Petrie's contention that the portraits were made before death there is no need to make a connection between the age of the sitter and his/her actual death, which might have occurred at a much later stage.

An experiment conducted in the 1990s by Richard Neave lends support to Petrie's view. Nieve used forensic evidence to reconstruct the faces of the two mummies, known as *The Two Brothers*, in Manchester Museum. He then compared the reconstructions to the portraits found in the coffins of the mummies and discovered a "striking resemblance".[85] Thus, while there are major points of continuity between the Hellenistic portrait of the dead and icons – the large expressive eyes, staring directly at the viewer, the gilding, the technique of tempera on encaustic (a wax-based painting medium), etc. – the presence in the image was fundamentally different. While most scholars would agree with Euphrosyne Doxiadis that there is "no doubt that portraits like those found in the Fayum are forebears of the icons"[86] there is a major difference in the terms on which the presence in the image has been predicated in each case.

The presence of the prototype in his image is a belief especially familiar from the cult of imperial portraits, which was universally spread throughout the late Roman period. As Alain Besançon points out "in the later Roman Empire, there was one icon throughout every commonwealth of the empire that indisputably and officially incarnated a god: the emperor's statue",[87] while Jas Elsner specifies that "right up to the early fifth century, high-quality imperial images [...] were being produced and used even when the empire had rejected a great number of the religious, administrative, and social tenets by which its earlier incarnations had

82 Ibid., pp. 150–1.

83 Ibid., pp. 144–5.

84 Cormack, *Painting the Soul*, p. 69.

85 Richard Neave, 'Richard Neave's Egyptian Encounter' in John Prag and Richard Neave, *Making Faces: Using Forensic and Archaeological Evidence* (College Station,TZ, 1997), p. 47; see Fig.5 on p. 48.

86 Doxiadis, *The Mysterious Fayum Portraits: Faces from Ancient Egypt*, p. 90.

87 Besançon, *The Forbidden Image: An Intellectual History of Iconoclasm*, p. 57.

lived".[88] In other words, even in the Christian period after theology had renounced the divinity of the emperor the cult of the imperial image, strictly speaking in contradiction with the position of the Church, persisted for a time. The fact that the image of Christ existed parallel to that of the "political god"[89] makes the question of the influences between the two of immediate relevance.

In Late Antiquity the portrait of the Emperor was treated as the Emperor himself and that was a practice defined by law.[90] It received the same honours or conversely, the same punishment as the person of a fallen emperor. Dio Cassius (*c.* AD 150–235) speaking of the end of Vitellius (d. AD 69) vividly describes how the wounded emperor and his statues were being dragged to prison.[91] Some of the imperial portraits could well be assumed to have held a physical similarity to the model. Others, however, obviously did not. An example is the porphyry group of the tetrarchs from *c.* AD 300 on the South façade of San Marco in Venice. With the four, almost indistinguishable figures, there is no attempt at individuation, which is in stark contrast to the Fayum portraits. In these and other imperial portraits physical similarity is a concern that has gone to the background in favour of a *type* which reveals broader political, ideological, and personal allegiances. Hadrian's (AD 117–138) portraits of the bearded emperor carry obvious associations with the type of the ancient philosopher. In this way, they stand in contrast, for example, to the military type of imperial portraiture (shortly cropped hair, deeply creased face, intense look) favoured by the soldier emperors at a later period. On the other hand, it is likely, as Elsner suggests, that Septimius Severus (AD 193–211), who had risen to power from the soldier ranks, too, sought to legitimize his power by propagating portraits of himself which displayed a marked resemblance to Marcus Aurelius (AD 161–180).[92] It seems that in the imperial portrait of Late Antiquity physical similarity is frequently forsaken for a portrait type which makes a statement. "Images of the emperor adopted a recognizable form", Elsner says, "even if their model (the emperor himself) might not always have been recognizable from them".[93] In other words, recognition was based on *type* disclosing an accepted form.

To the same artistic milieu belong divine images. In the period under our attention, i.e., Late Antiquity, the age-old notion that the image of a god can be understood as the god himself persisted. According to Elsner, "underpinning this ritual life of sacred images was the conviction that the statue did not just

88 Jas Elsner, *Imperial Rome and Christian Triumph* (Oxford and New York, 1998), p. 54.

89 Beançon, *The Forbidden Image: An Intellectual History of Iconoclasm*, p. 57.

90 Belting, *Likeness and Presence*, pp. 103 and 106.

91 Dio Cassius, *Dio's Roman History*, vol. 64 (London, 1954–1955), p. 255. The text reads: "Vitellius […] was dragged to the prison, as were also his statues, while many jests and many opprobrious remarks were made about them".

92 Elsner, *Imperial Rome and Christian Triumph*, p. 59.

93 Ibid., p. 54.

represent the deity but was – at least on some level – identical with the god".[94] Thus, the famous statue of Arthemis of Ephesus was "not just an image of Artemis or a particular interpretation of her, on the contrary, she *was* the goddess".[95] Artemidorus of Daldis (*c.* AD 96–180) bears witness to this attitude when he says in the *Oneirocritica* (The Interpretation of Dreams) that it is just the same whether one sees "Artemis herself [...] or her statue" in a dream, as "perishable statues" have "the same meaning as if the gods were appearing in the flesh".[96] At the same time, due to the lack of the model at hand, divine images rely even more than imperial ones on a recognizable type. Seeing the image of Artemis means, above all, being familiar with and recognizing the Artemis type.

Artistically the icon borrowed from all three groups of images, we mentioned – the portrait of the dead, imperial images and divine images. At the same time, it is the notion of a portrait-type, worked out by the latter two, which was particularly decisive in terms of the prototype – image relationship. Especially after Iconoclasm, a strict list of types was drawn for each saint and each subject. In this sense, Byzantine canonical art has its direct antecedent in divine and imperial portraiture of Late Antiquity. Presence – whether "full" or "partial" – was predicated not on physical resemblance, as with the majority of Fayum portraits, but on belonging to a recognizable type, sharing an accepted form.

Defining Presence in Pagan Platonism and Neoplatonism

Moshe Barasch draws attention to the frequently encountered motif in Greek literature, when the *psyche* (the human soul after it has left the body) acquires the characteristics of *eidolon* (image).[97] One of the author's examples is the famous episode in Homer's *Iliad*, when Patroclus's *eidolon/psyche* appears to Achilles. The *eidolon* is apparently the precise copy of the Greek hero. However, it lacks material substance, so when Achilles held out his arms to embrace his friend, "it vanishes like a wisp of smoke and went gibbering underground".[98]

The idea that the image is the same and yet different in some way from the original (here in terms of materiality) is expressed clearly by Odysseus, when he recounts his experiences in Hades:

---

94 Elsner, *Imperial Rome and Christian Triumph*, p. 205. See also Barasch's discussion of the animated image (Barasch, *Icon: Studies in the History of an Idea*, pp. 36–9).

95 Ibid.,p. 206.

96 Artemidorus Daldianus, *Das Traumbuch* (ed.) K. Blackertz (Munich, 1979), pp. 163–4.

97 Barasch, *Icon: Studies in the History of an Idea*, pp. 26–8. In Christian terminology, *eidolon* often means "idol" in contrast to *eikon* as "image", for example, in Origen's *Homily on Exodus 8*.

98 Homer, *Iliad*, XXII (Harmondsworth, 1950), p. 414.

After him I noticed Heracles in all his strength
A mere image, for himself (was) with the immortal gods.[99]

It is obvious how naturally this notion of only partial participation of the original in its image would fit in Plato's worldview, which presupposes different degrees of reality. In the *Timaeus* 52c and d, Plato discusses concretely the principle of the image as something real, the existence of which is derived:

> "Since that for which an image has come to be is not at all intrinsic to the image, which is invariably borne along to picture something else, it stands to reason that the image should therefore come to be *in* something else, somehow clinging to being, or else be nothing at all."

Consequently, Plato states that "as long as the one is distinct from the other in such a way that they at the same time become one and the same, and also two".

With the pagan Neoplatonists, the relationship between image and original continues along the lines intimated by Plato. The context is often the chain of emanations, as with Plotinus, which is also the chain of images, derived ultimately from the One. Each image resembles its original but only as a reflection. Its different position in the chain bears witness to its distinct reality from the one it reflects.

To describe the Neoplatonic understanding of the relationship between original and image Barasch proposes the category of resemblance. Resemblance here, however, is taken in its Antique connotation, which completely disregards the viewer's perception. The similarity between two items is decided on the basis of an inherent objective link between them, rather than on the accidental and subjective impression produced on the viewer.[100] It is in that sense that Proclus speaks of resemblance in the *Commentary to Parmenides* as being at once partial identity and partial otherness.

From all that has been said so far on classical Antique sources, dealing with the nature of the presence of the original in its image several conclusions can be drawn. First of all, the opinion of Barasch that there was in Antiquity and especially in late Antiquity a conceptual ambiguity on this issue seems acceptable. Certain texts bear witness to a full identification between prototype and image which I call "full presence". Other sources, some of the most outstanding of which belong to the Neoplatonic tradition, tell a different story. The image and the prototype are the same in so far as they resemble each other in an Antique and non-modern sense of the word, provided by an essential and objective link between them. At the same time they are different as they are not identical in certain aspects, as the two belong to a different reality. On those terms, the presence is "partial", i.e., the presence of the prototype in the image is real only as to what objectively connects them but

99 Homer, *Odyssey*, XI (Harmondsworth, 1946), pp. 601–602.

100 Barasch, *Icon: Studies in the History of an Idea*, p. 71.

should not be interpreted to imply that both belong to the same reality. It is possible to claim that God's image occupies a different reality than God himself, but, in some way, it partakes of divine reality. The fact that they occupy different realms does not sever the objective link between them. That there is such an objective link is the founding belief of the Byzantine theology of the icon.

Secondly, due to the conceptual ambiguity inherited from classical Antiquity and the lack of "an articulate, 'rational' theology of the holy image",[101] it remained for Byzantine authors to clarify the problem. The need for clarification became particularly urgent in the Iconoclast period but it had disturbed the minds of generations of Christians before that. While the theology of the icon, worked out during and after Iconoclasm, was an original contribution of Byzantine thought, it would have been impossible without the classical heritage. It can be maintained that writers like John of Damascus and Theodore the Studite took up the task of solving problems that had already emerged in Antiquity and especially in pagan Neoplatonism. As Barasch says, "there is no denying that the Iconoclast debate, especially the arguments put forward by the defenders of icons would be unthinkable without Proclus's heritage and what he represents".[102] At the same time, it is "equally clear that in some specific sense the Christian apologist of the sacred image had to start anew".[103]

Christian Sources. Byzantine Theology of the Image

Byzantine theology of the image was worked out during the Iconoclastic Controversy of the eighth and ninth centuries. In many ways, it addressed problems which had been in the centre of attention already in the early Christian period. In the first centuries after the adoption of Christianity the tension between full and partial presence of the prototype in the image continued. Whenever a full presence was assumed there was a real chance for the development of practices and beliefs rooted in magic, that go back to pagan Greco-Roman culture and beyond. In so far as the image we see *is* the prototype we are led to expect from it all the spiritual properties that belong to the prototype. It is one step from here to turning the image into an object of worship for its own sake. In this sense, it is only natural that Christian theologians would react strongly against attitudes reminiscent of idol worship.[104] Byzantine theology of the image undertook the task of clarifying the

---

[101] Ibid., p. 30. Hans Belting's view on this issue seems to be the same. According to him, especially in Late Antiquity, we keep coming across obviously "contradictory conceptions of the nature of the image" (Belting, *Likeness and Presence*, p. 27).

[102] Ibid., p. 86.

[103] Ibid.

[104] On the icon in early Christian thought see Barasch, *Icon: Studies in the History of an Idea*, pp. 93–183. The author discusses to some detail the early Christian apologists, Tertullian, Origen, Eusebius and Preudo-Dionysius. See also Gerhart Ladner, 'The Concept

conceptual distinction between idol and icon, which had piqued Christian thinkers for centuries. It was worked out, however, as a direct response to a strongly argued Iconoclast position. The iconoclastic arguments have to be reconstructed on the basis of indirect evidence or information supplied by the adversary group, i.e., the iconophiles, as there are almost no surviving primary sources by the iconoclasts themselves. They are important as they give an idea of the immediate background of the theology of the image.

The Iconoclastic Argumentation

The first argument that the Iconoclasts used is a Biblical one. There is an explicit ban on representational art in the Old Testament.[105] The Second Commandment is repeated throughout the text in several variants and most resolutely, in its visual implications, in Deut.4:15–18:

> "Therefore take good heed to yourselves. Since you saw no form on the day that the Lord spoke to you at Horeb out of the midst of the fire, beware lest you act corruptly by making a graven image for yourselves, in the form of any figure, the likeness of male or female, the likeness of any beast that is on earth, the likeness of any winged bird that flies in the air, the likeness of anything that creeps on the ground, the likeness of any fish that is in the water under the earth."

All such images should not be tolerated and should be actively destroyed. The Old Testament, therefore, urges that in cases of conquered nations: "But ye shall destroy their altars, break their images, and cut down their groves: [...] for the Lord, whose name is Jealous, is a Jealous God" (Exodus 34:12–14). At the same time, the prohibition of images goes alongside the observation that some images were permitted, even ordered, by God. For instance, Exodus 25:18–20 says: "And thou shalt make two cherubims of gold, of beaten work shalt thou make them, in the two ends of the mercy seat". Further, Moses is urged by his God: "make thee a fiery serpent, and set it upon a pole" (Numbers 21:8), an image that should have resembled in a curious way the pagan serpent of Aesculapius. The Iconophiles were not slow, of course, to draw attention to such texts from the Old Testament itself and exploit their implications.

The fact that the New Testament does not explicitly prohibit images was greatly exploited by both Iconophiles and Iconoclasts. However, we do come across contradictory statements, as in the Old Testament. St. John maintains that

of the Image in the Greek Fathers and the Byzantine Iconoclastic Controversy', *Dumbarton Oaks Papers*, 7 (1953): 1–34, reprint. Gerhart Ladner, *Images and Ideas in the Middle Ages: Selected Studies in History and Art*, vol. 1 (Rome, 1983), pp. 73–111.

105 On the Biblical prohibition of images see Barasch, *Icon: Studies in the History of an Idea*, pp. 13–22. There is useful bibliography on p. 21.

"no man hath seen God" (John 6:43–52), while later he reports Christ's saying that "He that hath seen me has seen the Father" (John 14:8–9).[106]

The second, philosophical, argument centres on the hierarchical opposition between matter and spirit going back to Plato. The theological concept of *theoprepres*, i.e., seemliness in regard to divine nature, belongs to the same discourse. The idea that what comes from man is not seemly of divine nature was given a logical interpretation by the Iconoclasts, who maintained that a figure made by human hands was *unseemly* in this sense of the word[107] and therefore God cannot be represented through matter and images fabricated by man.

Further not only is it unseemly to represent the Deity, but it is impossible to do so, as God, who is spirit lacks form. We can see how the idea that God is formless, especially prominent in Platonism, underlies almost all iconoclastic movements and ideologies in history.[108] According to Plato, while all things are conceived through Forms, thought by God, God himself is beyond Form. This conception famously reappears in Hegel's claim that "artistic form is purely finite, and therefore, incommensurate with infinite content which it is meant to represent".[109] Hence, Hegel's verdict that for the Greeks and the Christians, "God is not yet recognized as universal spirit",[110] unlike the Jews and the Muslims, for whom "spirit, as opposed to man, is seen as something devoid of form"[111] and for whom, consequently, there is "no place […] left for the visual".[112]

Hegel is relevant in this context, as he made explicit an idea that underlies the whole debate on images in Byzantium. The Iconoclastic Controversy was not about the value of art, but about *the possibility of representing the divine* and "on the issue of the image of God, Hegel is the only one since the iconoclastic controversy to have placed it at the centre".[113] When Hegel said that there was "no place for the visual" in Muslim and Jewish societies, he surely did not mean that Muslims and Jews completely lacked or were indifferent to visual images. In

---

106 It is noteworthy that the Koran, too, does not deal directly with the question of images. It is only in the Traditions about the life of Mohammed, probably written not before the eighth century, that the maker of images is described as "the enemy of God" (see extracts from the Traditions in Oleg Grabar, *The Formation of Islamic Art*, revised and enlarged edition (New Haven and London, 1987), pp. 82–3).

107 On the concept of *seemliness* see Besançon, *The Forbidden Image: An Intellectual History of Iconoclasm*, p. 20.

108 Besançon provides a useful outline of the philosophical arguments of Iconoclasm from the beginning, in Greek philosophy to Hegel and Kant (Besançon, *The Forbidden Image: An Intellectual History of Iconoclasm*).

109 Hegel, *Lectures on the Philosophy of World History Introduction*, tr. by H. Nisbet (Cambridge, 1975), p. 112.

110 Ibid., p. 111.

111 Ibid.

112 Ibid.

113 Besançon, *The Forbidden Image*, p. 203.

fact, he touched upon a very genuine problem which concerns the image of God, which was the pivotal problem for all religious iconoclasts in general. In this way, the Iconoclastic Emperor of Byzantium Constantine V Copronymus (718–775) got rid of all divine images covering the walls in the Milion of Constantinople and ordered representations of the six ecumenical councils and a lively depiction of the hippodrome games instead.[114] He did not seem to have been bothered that he was not following the Second Commandment to the letter, just as during the Reformation Calvin saw no contradiction in his position on images. No images at all were to be allowed in churches, according to Calvin, as God condemns "all shapes and pictures, and other symbols by which the superstitious imagine they can bring him near to them".[115] Calvin explicitly rejected the position on images of Nicaea II. He had the Byzantine Iconophiles and specifically Theodore the Studite in mind when he wrote: "their absurdities are so extreme that it is painful to even quote them".[116] At the same time, Calvin believed that "sculpture and painting are the gifts of God"[117] and he obviously saw it as completely fit for the artist to exercise his skill, so long as he/she did not pretend to represent God's image. Whether the rise of landscape painting and portraiture in Protestant countries is to be attributed to Calvin's direct influence is another matter, but he definitely allowed room for it.

The Iconophile Position

The simplest argument in favour of icons was the idea that they served an educational purpose. As St. John of Damascus says: "What the book is to the literate, the image is to the illiterate. Just as words speak to the ear, so the image speaks to the sight; it brings us understanding".[118] This notion ultimately goes back to St. Basil the Great,[119] but is mostly familiar from Pope Gregory the Great (590–604), whom John of Damascus cites on several occasions.[120] In his letter to Serenius, Bishop of Marseilles Pope Gregory states that: "the picture is for simple men what writing is to those who can read, because those who cannot read see and

114 In Judaism, it is the representation of God which is unacceptable on principle, while representational art as such has been tolerated to different degrees in different periods. Rabbi Akiba, for instance, explicitly authorized Jewish artists to practice their profession and even to make idols for the gentiles so long as, of course, they did not worship them (Besançon, *The Forbidden Image*, p. 74). Islamic legal texts contemporary with the Traditions allow the representation of human figures, so long as they are headless, while the Traditions themselves permitted figural representation in non-religious buildings.

115 Calvin, *Institutes of the Christian Religion*, Book I, Chapter II.

116 Ibid.

117 Ibid.

118 John of Damascus, *On the Divine Images*, p. 25.

119 Bychkov, *Vizantiiskaia estetika* (Byzantine Aesthetics), pp. 53, 79.

120 For example, in the First Apology, 11, p. 20; Third Apology III, 21, p. 77.

learn from pictures the model they should follow. Thus pictures are above all for the instruction of the people."[121]

A further task of the image is that it is the helpmate of writing in inducing remembrance, another doctrine associated with the name of Gregory the Great. The function of the Gospels and of the images which illustrated and supported the Gospels, was, in this sense, to serve as *memoria*. As Nicaea II decreed: "The more frequently they [the images] are seen in representational art, the more are those who see them drawn to remember and long for those who serve as models".[122] The retrospective orientation of the image (remembering) drew a bridge to the future as well (longing for). The image is meant to bring back the memory of a past moment of sacred history, and also to catch a glimpse of what was promised to come. The *memoria* as a device here properly functions simultaneously on these two levels – the one directed to the past and the other to the future – and it grows out of the larger concept of timeless eternity, referred to in the last chapter of this work. Thus, Pseudo-Dionysius, in connection to liturgy, talks of *anamnesia* or memory as remembering out of time.[123] What is "remembered" is the past and the future, as well as the present, a conception growing out of timelessness.

The Christological argument claims a role of the image as a confirmation of Christ's Incarnation. It is on this idea that the salvation of man rests and the visual bears a witness to it. Mankind is saved through Christ's consecration of the flesh and the material world. It is the mystery of God become man that makes it possible for St. John of Damascus to say: "Never will I cease to honour matter which wrought my salvation [...] Do not despise matter, for it is not despicable", "God has made nothing despicable"[124] and further, directly referring to the icon: "Therefore I boldly draw an image of the invisible God, not as invisible, but as having become visible for our sakes by partaking of the flesh and blood. I do not draw an image of the immortal Godhead, but I paint the image of God who became visible in the flesh".[125] The text of Nicaea II reads that "this [i.e., representational art] is quite in harmony with the history of the spread of the gospel, as it provides confirmation that the becoming man of the Word of God was real and not just imaginary and as it brings us a similar benefit".[126]

According to some authors, "the doctrine of images of the East rests completely upon Christology".[127] The strength of the Christological argument is that, even if not categorically refuting the Iconoclastic positions, it offers a viable alternative

---

[121] I am quoting from Wladyslaw Tatarkiewicz,*History of Aesthetics*, vol. 2: *Medieval Aesthetics* (The Hague and Warsaw, 1970), p. 104.

[122] Tanner (ed.), *Decrees of the Ecumenical Councils*, p. 136.

[123] Pseudo-Dionysius, 'Ecclesiastical Hierarchy' in his *Complete Works* (London, 1987), pp. 221-2.

[124] John of Damascus, *On the Divine Images*, pp. 23–4.

[125] Ibid., p. 16.

[126] Tanner (ed.), *Decrees of the Ecumenical Councils*, p. 135.

[127] Leeuw, *Sacred and Profane Beauty: the Holy in Art*, p. 315.

to them. It is based on a Biblical event (God's Incarnation) which sanctified matter once and for all, while at the same time, it changed the very nature of the divine (God, who is spirit, became flesh). What are more significant are the deeper implications of this view. To claim that the sacred image bestows grace would mean to compare it with the sacred word, which is unquestionably a means of grace and it is divinely inspired.[128] As St. Theodore the Studite said: "We should believe that divine grace is present in the icon of Christ and that it communicates sanctification to those who draw near with faith".[129] This was probably the final blow that the Iconophile theologians dealt on Iconoclasm. As Hans Asmussen passionately states: "Whoever has heard the message of the incarnation of the Word can never again pass by this form, can never again conceive of Christianity as formless ... Christianity is the Christ".[130] The implication is that religious art is made possible by the Incarnation. In Leeuw's opinion, "Christian theology does not begin with the creation, but with the redemption [...]. At this point begins also the theology of the arts."[131]

To accept the icon as a legitimate means of grace would involve proving that the image is a container of presence. In other words, the icon of Christ bestows grace so long as Christ is present in it in some manner. At the same time, the theology of the image should make it clear that the icon, unlike the idol, contains only partial and not full presence. It is, therefore, understandable that Byzantine theologians found it useful to refer back to Neoplatonic image theory which, as we saw, posited such a partial presence of the prototype in the image. This Neoplatonic framework is largely explained through Aristotelian terminology. Image and prototype were said to share *accident* but not *essence/substance* (see Glossary). Thus, form (*character*) as a synonym of *accident* becomes of primary importance. Gregory of Nyssa had reiterated this problem in the early Christian period by saying specifically in reference to the "eikon" that it is "the same in all respects to the prototype" and yet different from it "according to the properties of its nature".[132] Throughout St. John of Damascus's *Apologies*, this definition of the icon recurs on several occasions – an image "is of like character with the prototype, but with a certain difference", it is "not like the prototype in every way" (First Apology, p.19),[133] while St. Theodore the Studite says that "Christ is one thing and his image another by nature, although they have an identity (*tautotes*)" (First Refutation, p.31).[134] The dialectical nature of the argument becomes clear

128 A useful study on the image-word relationship, as well as on the role of the visual in Byzantine society, is Robin Cormack's *Writing in Gold* (London, 1985).

129 St. Theodore the Studite, *On the Holy Icons* (New York, Crestwood, 1981), p. 6.

130 Hans Asmussen, *Die Lehre vom Gottesdienst* (Munich, 1937). I am quoting from Leeuw, *Sacred and Profane Beauty: The Holy in Art*, p. 325.

131 Leeuw, *Sacred and Profane Beauty: the Holy in Art*, p. 328.

132 Migne (ed.), *Patrologiae Cursus Completu, Series Graeca*, 46, 41C.

133 St. John of Damascus, *On the Divine Images*.

134 St. Theodore the Studite, *On the Holy Icons* (Crestwood, New York, 1981).

when St. Theodore maintains that "if one says that the divinity is in the icon, he would not be wrong" and yet "the divinity is not present in them [the images] by a union of natures, for they are not deified flesh, but by a relative participation" (First Refutation, p.33). The nature of the identity is illustrated by the famous example of the impression left by the seal on wax (First Refutation, p.29), mentioned above. In his Third Refutation, Theodore the Studite states that "it is not the essence (*ousia*), but the form (*character*) of the prototype which is stamped upon it" (Third Refutation, p.103). Thus, in the case of the divine image, which is the main topic of St. Theodore's treatise, Christ "is seen mentally while (physically) absent" (Third Refutation, p.107).

What is rarely noticed is that there are actually two distinct assertions running through Byzantine theology of the image. On the one hand, word and image are supposed to have an equal status while, on the other hand, there is a recurrent implication that the image can do something that the word cannot and is an unique means of intuiting the Deity.[135] The special status of the image is a typically Neoplatonic idea, most often associated with Plotinus.[136] Through Plotinus and the Christian Neoplatonists this conception entered Byzantine theology and became an inseparable part of the Eastern Orthodox worldview. In Christian thought this idea was popularized by St. Dionysius the Areopagite (fifth century AD). It is exactly this philosophical tradition that connects the definition of the symbol with the idea of non-conceptual information which the image carries. In the Neoplatonic ontological hierarchy there are two means for the transfer of knowledge about the transcendent being – "the ineffable and mysterious on the one hand, the open and more evident on the other. The one resorts to symbolism and involves initiation. The other is philosophic and employs the method of demonstration" (Ep.IX.1105D).[137] "Those beings and those orders which are superior to us", says St. Dionysius in a characteristic passage, are "incorporeal" and "their hierarchy belongs to the domain of the conceptual and is something out of this world."[138] Our human hierarchy, on the contrary, we see filled with a multiplicity of visible symbols, through which we are led up hierarchically and according to our capacity

135 On the Iconophiles' belief in the distinct qualities of the visual image see, Barber, *Figure and Likeness: On the Limits of Representation in Byzantine Iconoclasm*, pp. 125–38. See also John Yiannias, 'A Re-examination of the 'Art Statute' in the Acts of Nicaea II', *Byzantinische Zeitschrift*, 80 (1987): 348–59.

136 See the already mentioned section in Plotinus, *Enn.*V.11.

137 Pseudo-Dionysius, 'Epistle IX' in his *The Complete Works* (London, 1987), p. 283. On this see Ronald Hathaway, *Hierarchy and the Definition of Order in the Letters of Pseudo-Dionysius* (The Hague, 1970), pp. 104–125. Epistle IX is widely accepted as a summary of the now lost 'Symbolic Theology' by Pseudo-Dionysius.

138 Pseudo-Dionysius, 'Ecclesiastical Hierarchy' in his *The Complete Works* (London, 1987), p. 197.

to the unified deification, to "the oneness and deifying simplicity of the Father".[139] Superior beings, as is appropriate to them, comprehend as pure intellects. We are led up, however, as far as possible, "by way of perceptible images"[140] to the contemplation of the divine. In this way, in fact, it is accepted that the symbol is inaccessible for human reason. At the same time, divine truth is contained exactly in a symbolic form. Consequently, the statement that the image is a symbol comes down to the claim that the image is the most adequate form, under which truth is revealed to man.

To summarize, what constituted the main theoretical problem for both Iconoclastic and Iconophile writers was the image of *God* rather than images in general. More precisely, for the Byzantines, as well as for later thinkers concerned with the problems of images, the central concern lay with the question if a visual representation can "contain" divine presence. If not, the image would be useless on theological grounds, whatever its aesthetic value. If, however, it could be proved that God, in some manner, is present in his image, this raises the value of the visual on a completely different level. The Icon-defenders undertook the task of developing a theory to counter Iconoclast arguments, relating mainly to the divine image. This meant that, in the long run, they elevated the image to the status of the word.

It is suggested, however, that the relationship between image and word, that the Iconophiles drew and elaborated on, remained paradoxical throughout the development of Byzantine theology. There is a constant tension between different, often contradictory, formulations of the status of the icon in relation to the word. It is sometimes necessary to read, as it were, between the lines and put statements into larger contexts. The idea that pictures can do something that words or any other medium cannot do was never worked out into a fully developed theory and it was actually veiled behind the word-image analogy. Probably the most convincing philosophical argument for the special status of the visual image is the one which stresses the role of symbolic image in acquiring non-conceptual knowledge, connected with the name of Pseudo-Dionysius. The full implications of this position were never fully explored and, in this sense, the re-sounding of the theme of the icon as symbol by Florensky, discussed in the following section, can contribute to a better understanding of problems, posed already in the medieval period.

The need to elaborate further on the implications of Byzantine theology of the image is felt particularly by those who believe that the Byzantine icon-defenders ultimately failed in their task. A line of scholars hold the view that both in theory and in practice sacred images remained subordinate to texts,[141] i.e., in this sense, the theology of the image failed in its assigned aim. The contemporary Russian scholar

---

139 Pseudo-Dionysius, 'Celestial Hierarchy' in his *The Complete Works* (London, 1987), p. 145.

140 Ibid.

141 Konrad Onasch, *Die Ikonenmalerei* (Leipzig, 1968), p. 191.

Victor Bychkov maintains that the icon-defenders have proved unable to produce a *clear conceptual proof* for the existence of "similar" images or the relationship between image and prototype. As he says, "the thesis that it is the hypostasis that can be represented and not the essence is little convincing".[142] Actually Bychkov suggests that Iconophile reasoning is based on the deep aesthetic impact of icons, but goes in the wrong direction, i.e., it does not try to explain the aesthetical factor but stays within the limits of the theological argument.[143] Alain Besançon, too, finds the Byzantine Iconophile arguments conceptually unconvincing. According to him, "the triumph of iconophilia was ambiguous, [...] secured by an unstable compromise, always on the point of splitting into two opposing factions, iconoclasm and iconolatry".[144] In particular, "the theological resolution of the problem, which entails a reaffirmation of the Incarnation, does not itself guarantee that the image expresses and realizes that goal of incarnation".[145]

The continued relevance of the Byzantine theology of the icon and particularly the theme of the presence of the prototype in the image will be illustrated through the writings of Florensky. The Russian author's understanding of the symbol comes once again to this theme, but this time the connection between the religious and aesthetic dimension of experiencing the image is much more explicit.

A Modern View: the Icon as Symbol in Florensky's Writings

It is worthwhile to pay attention to the affinity between the specific relationship between prototype and image in Byzantine theology, as outlined above and certain modern theories of the symbol. I will concentrate on Florensky's works on the symbol from the 1920s, which, I will suggest, have two main sources of inspiration. The more obvious reference is the Byzantine theology of the image, while the other is a much more recent one – German romantic philosophy – which, I believe to constitute an important aspect of the Russian author's work. By fitting the romantic definition of the symbol into an Eastern Orthodox theology of the image, the Russian writer, in effect, shifts the emphasis onto the visual implications of the symbol, which the romantics had studied mainly in relation to language. Genuine art, of which the ancient Russian icon is the supreme example, according to Florensky, is essentially symbolic.

The German romantic roots of Symbolist theory throughout Europe are well known. However, to my knowledge, they have not been remarked on in the case of Florensky. At the same time, Florensky's writings on the symbol have recently attracted attention mainly in the context of Florensky's relations with Russian

142 Bychkov, *Vizantiiskaia estetika* (Byzantine Aesthetics), p. 128; my translation.

143 Ibid., p. 129 This, actually, is the starting ground for Bychkov's own study of Byzantine aesthetics.

144 Besançon, *The Forbidden Image: An Intellectual History of Iconoclasm*, p. 3.

145 Ibid.

Symbolists.[146] And for the Symbolists, the German romantics were inspirational – both directly and through Nietzsche's reading. No one was closer to Florensky, both personally and ideologically, than the Symbolist poet and main theoretician of the movement Viacheslav Ivanov. I have discussed elsewhere, in some detail, the affinities between Ivanov's and Florensky's respective theories of the symbol and their common grounding in German romantic theory.[147]

Even though writing at a different historical period, Florensky found himself, at least in some ways, in a situation similar to the one facing the Byzantine icon-defenders in the eighth and ninth centuries. Russian intellectuals since the nineteenth century had been reacting against what they felt was a modern variety of iconoclastic attitudes. Florensky is not, by any means, the only one who pointed at "the present variations on this theme in the counter-arguments to Protestants and of rationalism".[148] Dostoevsky's character Fyodor Karamazov, who spits at the icon of his wife and teases her, saying: "You see that icon of yours? … You think it can work miracles, but I'll spit on it in your presence and nothing at all will happen to me!",[149] becomes an exponent of the iconoclastic tendencies of the modern age. At the same time, writing on "the eve of the new iconoclasm"[150] following the October Revolution, Florensky was no longer concerned with distinguishing the Christian image from the pagan idol, as Byzantine theologians had been. His immediate interest was to stress the sacred dimension of the icon, which was glorified as a symbol of Russian and Eastern Orthodox identity. Thus, Florensky identifies himself with a line of Russian thought concerned with the

---

146 On Florensky's relations with the symbolists see Elena V. Ivanova (ed.), *Pavel Florensky i simvolisti* (Pavel Florensky and the symbolists) (Moscow, 2004). Florensky's correspondence with Bely has been published in Ivanova's book, while the correspondence with Viacheslav Ivanov is in 'Arkhivnie materiali i issledovanii' in *Russkoe Slovo* (Moscow, 1999), 93–7. Florensky contributed to several prominent Symbolist journals – his first article, 'On Superstition', came out in *Novii put'*, the journal published by Dmitrii Merezkovsky and Zinaida Hippius, while he also published in Valerii Briusov's *Vesy.*

147 See my "'Beauty Will Save The World': The Revival of Romantic Theories of the Symbol in Pavel Florenskii's Works", especially the section on "Florenskii's View on the Symbol within the Context of Russian Symbolism: Florenskii and Viacheslav Ivanov", *Slavonica*, 14/1 (2008): 49–51.

148 Pavel Florensky, 'Ikonostas' (Iconostasis) in Pavel Florensky, *Khristianstvo i kultura* (Christianity and Culture) (Moscow, 2001), p. 549; the translation is mine; the English translation in Pavel Florensky, *Iconostasis* (Crestwood, New York, 1996, reprint, 2000), p. 70 is imprecise.

149 Fyodor Dostoevsky, *The Brothers Karamazov*, vol. 1 (Moscow, 1980), p. 210. Actually, in Russian the word that Dostoevsky uses is "obraz" (image) for "icon" in the English translation (Dostoevsky, *Bratia Karamazovi* (The Brothers Karamazov) in Fyodor Dostoevsky, *Polnoe sobranie sochinenii* (Complete Works), Leningrad, vol.14, p. 126).

150 Alexander Mikhailov, 'O. Pavel Florensky kak filosof granitzi' (Father Pavel Florensky as a philosopher of the boundary), *Voprosy izkustvoznania* (Questions of the Theory of Art), 4 (1994): 60, the translation is mine.

problem of iconoclasm, conceived both broadly and with regard to the Russian sacred icon in particular, which was rediscovered at the time.[151]

It is possible to see Florensky's essay "The Church Ritual as a Synthesis of the Arts' (1918) as a direct response to Lenin's Plan for Monumental Propaganda of the same year. The Plan was explicitly concerned with the demolition of tsarist monuments and the erection of new ones, glorifying the new regime. It further stipulated the removal of objects of historical and artistic value from their original locations in churches and monasteries to the newly-founded public museums and art galleries. The overt purpose was preservation but it went along with the intentional destruction of the original meaning of religious objects. Florensky's essay called for keeping the Lavra of the Trinity in Sergiev Posad, a monastery founded in the fourteenth century by the country's patron saint, Sergius of Radonezd, in its original role.[152] The Lavra, Florensky maintained, was "infinitely necessary to Russia".[153]

It is important to note the immediate historical background of Florensky's works on the symbol. At the same time, the Russian writer's response to early Soviet iconoclasm is part of his much larger and more general concern with the crisis of modernity, a concern he shared with a host of European thinkers at the turn of the twentieth century and later. In this intellectual context, it was natural to create a link with the German romantic philosophers, who were among the first to pay systematic attention to the dangers inherent in the culture of the modern age.

The symbol being one of the most allusive cultural categories, the Russian author takes care to specify which of its uses he was adhering to. A main concern of Florensky, in the *Iconostasis* and other writings,[154] has been to distance himself from certain interpretations of the symbol which tend to sever the connection between prototype and image or, to use Sausseurian terminology, the signifier and

---

151 See, for example, Evgeny Trubetskoy, *Umozrenie v kraskakh: tri ocherka o russkoi ikone* (Conteplation in Colours: Three Essays on the Russian Icon) (Paris, 1963, first in 1915–1918). The revived interest in Russian medieval art dated back to the middle of the nineteenth century.

152 I have discussed these developments in some detail in my 'Re-contextualizing Holy Images in Early Soviet Russia: Florensky's Response to Lenin's Plan for Monumental Propaganda' in *Der Sturm der Bilder/The Clash of Images*, published for the University of Hamburg, forthcoming.

153 Pavel Florensky, 'The Church Ritual as a Synthesis of the Arts' in Pavel Florensky, *Beyond Vision: Essays on the Perception of Art* (ed.) N. Misler (London, 2002), pp. 110–111.

154 See Elena Nekrasova, 'Neosushtestvennii zamisel 1920-h godov sozdania "Symbolarium" (Slovaria simbolov) i ego pervii vipusk "Tochka"' (The unrealized project of the 1920s for the creation of 'Symbolarium' (Dictionary of Symbols) and its first issue 'Point" in *Pamiatniki kulturi. Novie otkritia* (Monuments of Culture. New Discoveries), annual publication, Leningrad, 1994, pp. 99–115. The text of the first issue is on pp. 100–115.

the signified. Florensky distinguishes between "two thresholds of receptivity"[155] of the symbol – an "upper" one, at which the symbol preserves some identity with the prototype, as in its ancient usage, and the "lower", at which the ontological connection between the two entities has been broken, as in modern times.[156] Florensky's own understanding of the symbol apparently belongs to the former, "upper threshold".

The vertical direction, intimated by Florensky's "upper" – "lower threshold" is evident in Viacheslav Ivanov's much better known and earlier *liniia voskhozhdeniia* (line of ascent) – *liniia niskhozhdeniia* (line of descent), which describe the process of artistic creation and are again connected to the theme of the "boundary.[157] Both Florensky's and Ivanov's terminology and meaning, I believe, derive from the romantic opposition between symbol and allegory. It is known that in the late nineteenth and early twentieth centuries there was a reawakened interest and even vogue for romantic theories of the symbol, of mythology, of language and other related fields. Friedrich Vischer's essay "Das Symbol" (1887)[158] became particularly influential.[159] Vischer's conception of the two polarities of the magical-associative and the logical-dissociative was directly inspired by the romantic opposition between symbol and allegory.[160] Florensky himself, without mentioning the romantic source of the distinction specifically in the *Iconostasis*,[161] contrasts the "allegorized symbol" to "true symbols" (*nastoiashchie simvoli*).[162]

---

155 Pavel Florensky, cycle of lectures on 'Cult and Culture'. For the English translation, I am citing from Viktor Bychkov, *The Aesthetic Face of Being: Art in the Theology of Pavel Florensky* (Crestwood, New York, 1993), p. 7.

156 It is significant that Florensky associates the latter with modern times in general and so, identifies it as an expression of the crisis of modernity.

157 On the "ascent-descent" dichotomy in Ivanov's poetry, see Victor Terras, 'The Aesthetic Categories of *Ascent* and *Descent* in the Poetry of Viacheslav Ivanov', *Russian Poetics*, 1975.

158 Vischer, Friedrich, 'Das Symbol' in Friedrich Vischer (ed.), *Philosophische Aufsätze, Eduard Zeller zu seinem 50 jahrigen Doctor-Jubiläum gewidmet* (Leipzig, 1887), pp. 153-93.

159 For example, according to Edgar Wind F.Vischer's work played a formative role in Aby Warburg's views of symbolism (Edgar Wind, 'Warburg's Concept of *Kulturwissenschaft* and Its Meaning for Aesthetics'(1931) in his *The Eloquence of Symbols: Studies in Humanistic Art* (Oxford, 1950, reprint,1983), p. 27). See also Matthew Rampley, 'From Symbol to Allegory: Warburg's Theory of Art', *Art Bulletin*, LXXIX/11 (1997): 41–57.

160 For a useful account of the romantic theory of the symbol, see Tzvetan Todorov, *Theories of the Symbol* (Oxford, 1982, first in French in 1977).

161 Schelling is mentioned in another work as one of the sources of *sobornost* (all-unity), a concept of great importance in Russian religious philosophy (Florensky, *Sobranie sochineniia* (Complete Works), vol. 2 (Paris, 1985), p. 295), while Florensky directly borrows the German philosopher's term of "concrete idealism" to describe his own world-view" (Florensky, Ibid., vol. 3, pp. 39–40).

162 Florensky, 'Ikonostas' (Iconostasis), pp. 560–1; Florensky, *Iconostasis*, p. 85.

Russian art from the end of the sixteenth century is said to exhibit "the spirit of allegory", while earlier art of the fourteenth and fifteenth centuries is symbolic. To grasp fully Florensky's meaning we have to go back to the romantic opposition between symbol and allegory.

The first writer to draw the opposition between symbol and allegory is the art historian Heinrich Meyer (1760–1832) in his essay *Uber die Gegenständer der bildenden Kunst* (On the Objects of the Plastic Arts) (1797).[163] In his notes for the edition of Winckelmann's *Werke* Meyer gives the following definition of the symbol-allegory pair: "Symbolic representation is the general concept itself, rendered perceptible; allegorical representation signifies only a general concept different from itself."[164] The opposition is drawn on the basis of a difference in the case of allegory versus a sort of identity, or "transitivity", as Tzvetan Todorov says,[165] in the case of the symbol. In other words, the organic link between allegory and what it represents is severed, as in Florensky's "lower threshold of receptivity", while it is preserved with the symbol, at the "upper threshold". In another text, Meyer elaborates on the function of symbolic visual representation of "divinities", in which "figurative art [...] forces ideas and concepts themselves to make their appearance in a perceptible way, it requires them to enter into space, to take shape and present themselves to the eye".[166]

In his lectures on the *Philosophy of Art* of 1802–1803, Schelling goes back to this definition of the symbol. He makes even clearer that one of the fundamental characteristics of the symbol is the fusion of the general, i.e., ideas and concepts and the particular, i.e., the appearance of ideas and concepts "in a perceptible way", as Meyer had put it. Thus, "a picture is symbolic", according to Schelling, "whose object not only signifies or means the idea but is *itself the idea*".[167] Throughout his passages dealing with the symbol, Schelling tirelessly insists on the unity of general and particular, understood in the sense here, i.e., the symbol in its particular representation of a general notion is simultaneously that notion.

We have to bear in mind both strands of thought – the Byzantine and the romantic – when considering Florensky's examples. There is a complex interplay of meanings, burdening the symbol as image in passages as "Now I look at an icon and I say to myself: 'Behold, this is She – not her picture but She Herself, contemplated by means of, with the aid of, iconographic art. As through a window I see the Mother of God!",[168] and the paragraph, concerning Egyptian funerary

---

[163] Goethe, a friend of Meyer's, wrote an essay under the same title that same year, but it was published only later.

[164] The English translation is from Todorov, *Theories of the Symbol*, p. 214.

[165] Todorov, *Theories of the Symbol*.

[166] Ibid., p. 213.

[167] Friedrich Schelling, *Philosophy of Art* (Minneapolis, 1989, first in German in 1859), p. 151.

[168] Florensky, P., 'Ikonostas' (Iconostasis), p. 548. The English translation here is from Bychkov, *The Aesthetic Face of Being: Art in the Theology of Pavel Florensky*, p. 80,

art from the Roman period which is an antecedent to the icon.[169] Referring to the Fayum portraits of the dead, Florensky says that they were the object of elaborate devotional practices, based on the belief that the deceased is present in his portrait. The relative or friend of the dead would say "This *is* my father, brother, friend" (the italics are mine), and not "This is the paint on my father's face" or "That's the mask of my friend", etc.[170]

The identity that Florensky claims between prototype and image clearly takes us back to the Neoplatonic and Byzantine understanding of the image which we discussed earlier. The ancient roots of this definition of the symbol are well realized by the romantics themselves and are repeatedly acknowledged, especially in their writings on mythology. For Schelling, the right approach to mythology starts from the realization that it is "universal symbolism or universal *representation* of *ideas* as real".[171] In this sense, "in allegory the particular merely *means* or *signifies* the universal; in mythology it itself *is* simultaneously also the universal".[172] This is why we must not say, for example, that "Jupiter or Minerva *means* or *signifies* this or is supposed to signify it … They do not *signify* it, they *are* it themselves".[173] In the same fashion, an icon of the Mother of God, as in Florensky's example, does not merely signify its prototype but is, in the sense here, the Mother of God.

The symbol, therefore, by displaying an ontological identity between prototype and image opposes a tendency, evident with the allegory, which disconnects the link between the two entities. It is this view that underlies Florensky's view of symbolism. As the Russian writer says: "I acquired the basic thought of my worldview: that what is named in name, what is symbolised in the symbol, the reality of what is pictured in the picture, is indeed present, and therefore the symbol *is* symbolised".[174]

To discover the function that the symbol plays in Florensky's worldview entails going back to the theme of the boundary. As we saw in Chapter 1, Florensky discusses specifically dreams and mental (and by implication, artistic) images as belonging to the boundary zone between the two worlds. The Russian word that Florensky uses for mental image is "lik", which can be translated as "countenance" and is used in contrast to "litzo", which means literally "face". The distinction is quite important in the course of the discussion, even though it is difficult to

---

which I find better than the translation in Florensky, *Iconostasis*, p. 69.

169 Actually, as was mentioned earlier, the origins of the icon are much more complicated.

170 Ibid., p. 625; the translation is mine; the English translation in Florensky, *Iconostasis*, is on p. 164.

171 Friedrich Schelling, *Philosophy of Art*, p. 17.

172 Ibid., 47.

173 Ibid., p. 42.

174 Florensky, 'Ikonostas' (Iconostasis), p. 625; Florensky, *Iconostasis*, p. 164.

express in English. The "lik" is "an ontological *gift* of God",[175] it is an expression of "the possibility that God's image [...] be embodied in life, in the personality and thus to present it through the "litzo".[176] The authentic icon is, in this sense, exactly a "litzo", which reveals the "lik". The most important for our purposes is that both dreams and mental (and, by implication, as long as they are authentic, artistic) images, understood in the way just mentioned, are symbols. What is said about dreams is valid about images in general – in them "*both* shores of existence are given to the consciousness" "simultaneously but with differing orders of clarity".[177] The iconic images become "the visible witnesses of the invisible world";[178] they "dwell simultaneously in two worlds, combining within themselves the life here and the life there".[179]

The boundary is the area of being between the two worlds – it touches both of them and, at the same time, coincides with neither. That is, the boundary is something else, different from the other two, but having something in common with both of them. The symbol belongs to the boundary and, in this sense, it is the link between the worlds and possesses characteristics of both the transcendent and the immanent worlds. This idea is expressed in the following definition by Florensky:

> "A being that is greater than itself – this is the basic definition of the symbol. A symbol is something that manifests in itself that which is not itself, that which is greater than itself and is nevertheless manifested through itself [...] a symbol is an essence energy of which is joined, or, more precisely, commingled, with the energy of another essence, more worthy in a given respect, and which thereby carries this other essence in itself."[180]

The iconic image possesses the characteristics of the symbol, understood in the sense here. Its essence is to make present and accessible to the senses (a characteristic of this world) of what is, in principle, invisible and only spirit (a characteristic of the world beyond). The authentic icon, which fulfils its purpose, incorporates both these moments, as in Florensky's example of the image of the Mother of God, which is "not Her picture, but She Herself". At the same time,

---

[175] Ibid., p. 535; the translation is mine; there is nothing to suggest in the Russian text the strange phrase "ontologically actual" in the English translation in Florensky, *Iconostasis*, p. 51.

[176] Ibid; the translation is mine.

[177] Ibid., pp. 529–30; Florensky, *Iconostasis*, p. 43.

[178] Ibid., 542; Florensky, *Iconostasis*, p. 60.

[179] Ibid, p. 542; ibid., p. 62.

[180] Pavel Florensky, 'Imeslavie kak filosofskaia predposilka' (Onomadoxy as a Philosophical Premise) in Florensky, P., *Khristianstvo i kultura* (Christianity and Culture) (ed.) A. Filolenko, Moscow, 2001), p. 279. The English translation is from Bychkov, *The Aesthetic Face of Being: Art in the Theology of Pavel Florensky*, p. 70.

the icon "on its own" – i.e., apart from the spiritual vision – "is neither an image nor an icon, but a wooden board".[181] That is, the iconic and the symbolic of the image are in its connection to the prototype. Florensky illustrates this notion in a clear and straightforward way. The window on its own is no more than "wood and glass",[182] but once we are able to see the light through it the window becomes "that very light itself" and is not just "'like' the light".[183] In the same way, we have to understand the other example about the friend or relative, looking at the portrait of the deceased and saying: "This is my father, brother, friend". Thus, Florensky's understanding of the symbol organically belongs to the Eastern Orthodox tradition which stresses the idea of the simultaneous existence of both, from the point of view of reason, contradictory aspects of the icon. At the same time, this view of the symbol is very pronounced with some romantic philosophers.

What Todorov refers to as "syntheticism"[184] or the fusion of contraries – matter and spirit, infinite and finite, etc. – is a constant feature of romantic philosophy. The romantics, however, made systematic and explicit the aesthetic dimension of this fusion and this is of special significance for the purposes of the present text. Schelling again comes to mind with his *System of Transcendental Idealism* (1800). Tellingly, Schelling's system culminates with art which has the property of absorbing all contraries. As Schelling says, "every aesthetic production [...] is an infinite finitely displayed".[185] He immediately continues that "the infinite finitely displayed is beauty".[186] August Schlegel (1767–1845) takes on Schelling's statement in the context of the discussion of the symbol:

> "According to Schelling, the infinite represented in finite fashion is beauty [...] I am in full agreement on this point, I should simply prefer to formulate this expression as follows: the beautiful is a symbolic representation of the infinite; for in this way we can see clearly how the infinite can appear in the finite [...] How can the infinite be drawn to the surface, made to appear? Only symbolically, in images and signs [...] Making poetry (in the broadest sense of the poetic that is at the root of all arts) is nothing other than an eternal symbolizing."[187]

---

181 Florensky, 'Iconostas' (Iconostasis), p. 545; the translation is mine; the English translation in Florensky, *Iconostasis* is on p. 65.

182 Ibid.; the translation is mine; the English translation in Florensky, *Iconostasis* is on p. 65.

183 Ibid.; Florensky, *Iconostasis*, p. 65.

184 Todorov, *Theories of the Symbol*, p. 184.

185 Friedrich Schelling, *System of Transcendetal Idealism* (Charlottesville, 1978, first in German in 1800), p. 225.

186 Ibid.

187 August Schlegel, *Vorlesunger über schöne Literatur und Kunst*, vol. 1: *Die Kunstlehre* (Theory of Art), (Stuttgard, 1963); the English translation is from Todorov, *Theories of the Symbol*, p. 198.

The above passage contains, in a condensed form, all the main features of Eastern Orthodox aesthetics. On the one hand, it makes explicit notions that often lie dormant in the Byzantine theology of the image. On the other, it provides the starting ground for the understanding of the icon in Russian philosophy and specifically in Florensky's thought. Still further, it outlines the background of a line of twentieth-century critique of modernity. The main thrust of the argument is organized around the definition of the symbol as "the infinite represented in a finite fashion". In other words, the symbol is the supreme paradox, containing the greatest possible contradiction. We saw that this idea was at the heart of the Byzantine view of sacred images. When Florensky describes the symbol as "a being that is greater than itself" and "manifests in itself that which is not itself", I believe, he keeps to the Eastern Orthodox understanding but via the romantic definition of the symbol.

The romantics felt the dangers of the allegorical view, but they also offered a way out by burdening art with a redeeming role in a disintegrating modern culture. Florensky takes up this idea and interprets it in an Eastern Orthodox context. It is the medieval Russian icon that has the ability to restore the lost "magical" identity between "being" and "thing" which lies at the heart of art. Thus, while echoes of Nietzsche's *The Birth of Tragedy* (1872)[188] can be easily heard in Florensky's view of art, with the Russian author the visual acquires a metaphysical significance which is both non-Platonic and non-Nietzschean. In an obvious allusion to Nietzsche, Florensky defines the symbolic image as one which reflects the soul's Dionysiac experience of "sundering of the bonds of the invisible".[189] The "symbolic images"[190] are revealed to the soul only after "soaring up to the invisible, the soul descends again into the visible".[191] This, Florensky says, "is the Apollonian perspective on the spiritual world", i.e., the Apollonian dimension of the icon as a visual art becomes a vehicle for disclosing its Dionysiac essence, in a specifically Christian context. So, while for Nietzsche it is the art of music (and, by implication, tragic drama which can be set to music) that represents the Dionysiac *par excellence*, Florensky burdens the icon with Dionysiac qualities. At the same time, it is interesting to notice that Florensky's insistence that it is the *Russian* sacred image and *Russian* culture that have this significance makes yet another contact with Nietzsche, who in various works claimed that a source of

[188] *The Birth of Tragedy* was often the first book by Nietzsche Russians would read (see Bernice Rosenthal, 'Introduction', in Bernice Rosenthal (ed.), *Nietzsche in Russia* (Princeton, 1986), p. 10).

[189] Florensky, 'Iconostas' (Iconostasis), p. 531; the translation is mine; Florensky, *Iconostasis* is on p. 45.

[190] There does not seem to be anything in the Russian text to justify the English translation "real *appearances*". The Russian phrase is "simvolicheskie obrazi", which is literally "symbolic images".

[191] Ibid.

hope for the future lay exactly with Russia.[192] Very probably, a direct influence on both authors in this context was Dostoevsky.[193]

The discourse on the role of art continues in the later twentieth century. It focuses very much on the idea of modernity's loss of immediacy and genuine contact with being – for instance, in Walter Benjamin's notion of the loss of aura in a technological age, in the late Heidegger's writings on art, in Theodore Adorno's thesis that the loss of the name as a primary constituent of language is an indicator of the alienations of modernity,[194] etc. All these writings converge on the thesis that modern culture is in a state of crisis and art is emblematic of this crisis. In one way or another, they are playing on Hegel's idea of the end of art. As a result of art's becoming aesthetical it has been alienated from truth. Jay Bernstein has called this "aesthetic alienation". But if we understand aesthetics as "the theoretical discourse that comprehends art in its autonomous, post-Christian guise",[195] i.e., as Kantian aesthetics, and, if we want to find a way out of the predicament we have reached, it makes sense to reach back to pre-modern views of art. Reviving the notion of the symbolic role of art represents such an attempt.

This is what Heidegger does when he discusses art explicitly in terms of the symbol. Almost certainly never having heard of Florensky's work, it is interesting that even Heidegger's highly specific language sounds at times close to the Russian thinker's. Surely the statement that "the work is a symbol"[196] in the sense that it is "a thing, but a thing to which something else adheres"[197] brings to mind Florensky's definition as "a being greater than itself". That the sculpture of a God "is not a portrait whose purpose is to make it easier to realize how the god looks, rather, it is a work that lets the god himself be present and thus *is* the god himself"[198] is another way of insisting, as Florensky had done, that in the icon we see the Mother of God, "not her picture". Florensky would have readily agreed with Heidegger's claim that "the work is a work, as long as god has not fled from

---

192 Susan Ray, 'Afterword: Nietzsche's View of Russia and the Russians', in Rosenthal (ed.), *Nietzsche in Russia*, pp. 393–401.

193 Nietzsche had read Dostoevsky in French translation. The affinities between certain views of the two thinkers have been noted early on, for instance in the 1903 study on *Dostoevsky and Nietzsche: A Philosophy of Tragedy* (in Russian) by Lev Shestov, another prominent representative of the Russian Religious Renaissance. On this, see 289–302.

194 Theodor Adorno and Max Horkheimer, *Dialectic of Enlightenment* (London, 1986).

195 Jay Bernstein, *The Fate of Art: Aesthetic Alienation from Kant to Derrida and Adorno* (Cambridge, 1992), p. 1.

196 Martin Heidegger, *Poetry, Language and Thought* (New York, 1971), p. 40.

197 Ibid., p. 20.

198 Ibid.

it"[199] and that art serves to "accomplish being"[200] and "unconcealedness".[201] Once again, though, Heidegger, as the romantics before him, sees the symbol at work in classical Greek art,[202] while for Florensky it is the medieval Russian icon that is ultimately symbolic.

In Florensky's writings the view of the symbol as displaying the being/presence of the prototype carries romantic overtones, as well as the more obvious medieval ones. The Russian author's understanding of the symbol is not only an attempt to keep to the spirit of Eastern Orthodox theology of the image, but is part of a concern about the modern condition in a world in which, as Heidegger put it, "all distances in time and space are shrinking" and yet "the nearness of things remains absent".[203] Even though not fully developed, Florensky's understanding of the symbol is interesting for bringing together and re-sounding these two trends of thought – the Byzantine and the romantic – in view of a constructive approach to the crisis of modernity.

Conclusion and Implications

In this chapter it has been proposed that pictorial form can act as a means of signalling the presence of the prototype in the image. If we accept this interpretation, we would be committed to the view that the icon is a container of presence only in a partial manner, i.e., the identity between prototype and image is in certain but not all aspects. More concretely, there is an identity of form but not of essence. This position on the image was first worked out in Late Antiquity by pagan and Christian Neoplatonists and was further developed by Byzantine writers of the Iconoclast period.

A major obstacle posed by the theory of the image, outlined here are the very terms on which it has been postulated. To grasp the meaning of issues as presence, resemblance and form we need to adopt a special perspective – that of the paradox which lies at the heart of Christianity. As Kierkegaard has showed, this is a challenging task. This explains the enduring tension between what Jill Dubisch calls "religion as prescribed" and "religion as practiced".[204] While the official position of the Church accepts an elite theological formulation of "partial real

199 Ibid., p. 43.

200 Ibid., p. 138.

201 Ibid., p. 56.

202 In keeping to his overall position, Heidegger attempts consistently to describe the Greek temple in non-aesthetical terms. On this see Gerald Bruns, *Heidegger's Estrangements: Language, Truth and Poetry in the Later Writings* (New haven and London, 1989) for an outline of Heidegger's terminology in his works on art.

203 Ibid., 165.

204 Jill Dubisch, 'In a Different Place' in Ellen Badone (ed.), *Religious Orthodoxy and Popular Faith in European Society* (Princeton, c.1990), pp. 113-39.

presence" of the prototype in the image, in popular usages images are frequently almost automatically endowed with "full real presence".[205]

A further problem arises from the realization that the notion of presence makes the framework of Kantian aesthetics inadequate for the analysis of icon art. It is obvious that if we see the image as an animated image, there are important implications relating to some of our main assumptions about art, representation and aesthetic experience.

First and foremost, we can no longer assume a "disinterested" aesthetic experience, which has been one of the most deeply embedded notions since Kant. A beholder facing a divine presence in a representation can in no way be expected to be "indifferent [...] to the real existence of the object of this representation".[206] Neither can he/she be assumed to do no more than "play the part of judge in matters of taste"[207] as a result of this indifference and disinterestedness. At the same time, "leaving out of account the existence of the thing is precisely the characteristic and essential reality of aesthetic representation".[208]

Second, the very terms on which visual experience has been postulated are questioned. In order to have visual experience, we need a seeing subject contemplating the seen object. Since at least Kepler[209] and later in Cartesian philosophy, this has been interpreted as a split between subject and object. The problem is if this modern aesthetic experience – a distanced, "disinterested" and neutral viewing on part of the subject of an object that exists apart from him/her – can describe adequately what happens in the interaction between image and viewer in a religious context. The experience of the subject contemplating an animated image seems to have been revolved not only round the act of seeing but more significantly, around *being seen*. The image that "stares back"[210] can hardly be described as an object of contemplation in the narrow sense of the term. It seems that a much more intimate and active interaction took place when the viewer was faced with a *re-presentation* of the deity. Bruno Latour has made a useful distinction between re-presentation and representation in his thesis that in the Renaissance the latter mode superseded the former[211] or, in other words, the image

---

205 This distinction is made clear, for instance, in Cormack's interesting account of the reactions by high-minded theologians and the average laymen to the Turin Shroud (Cormack, *Painting the Soul*, pp. 115-26).

206 Kant, *Critique of Aesthetic Judgement*, tr. by J. C. Meredith (Oxford, 1911), p. 43.

207 *op. cit.*

208 Ernst Cassirer, *Kant's Life and Thought* (New Haven and London, 1981), p. 311.

209 Kepler's discovery of the retinal image implied, among other things, a notion of the role of the eye as more neutral and less active in the process of vision.

210 James Elkins, *The Object Stares Back: On the Nature of Seeing* (New York and London, c.1996).

211 Bruno Latour 'Opening One Eye While Closing the Other ... A Note on Some Religious Paintings' in Gordon Fife and John Law (eds), *Picturing Power: Visual Depiction*

was no longer seen as a container of presence. In this process, according to Latour, the heavens were turned into a sky since, to invert Heidegger's expression,[212] the god had fled from his image.

Some of these questions were addressed directly or at least implied in Florensky's works on the symbol from 1920s. I have suggested that Florensky revives the Late Antique and Byzantine understanding of the image on the background of themes already familiar from nineteenth century thought. At the same time, the understanding of the image that the Russian author subscribes to shows points of contact with ideas voiced by a line of thought in the twentieth century and to the present – for example, in phenomenology[213] and psycho-analysis[214] – which has questioned the principles of visuality, implied by Kantian aesthetics. It seems to me that modern studies could be enriched by an insight into the nature of the icon, which can serve as providing an alternative mode of visuality to the dominant, epistemological one.

---

and Social Relations (London, 1988).

[212] See above, the section on Florensky in this chapter.

[213] The ontologogical vision implied by the antique image is a recurrent theme, underlying Husserl's "phenomenological seeing", Heidegger's "*aletheia*", i.e., "unhiddenness", Merleau-Ponty's notion of embodied vision. Merleau-Ponty's remark that "the world is all around me, not in front of me" (Maurice Merleau-Ponty, 'Eye and Mind' in *The Primacy of Perception* (Evanston, Ill., 1964), p. 178) refers exactly to the mingling of the viewer with the world on view. On visuality in phenomenology see Chapter 5 in Martin Jay, *Downcast Eyes: Denigration of Vision in Twentieth Century French Thought* (Berkeley, 1993).

[214] See Lacan's notion of the "gaze" versus the "eye". His concept of the "mirror stage" is meant, above all, to go beyond disinterested observation as evident in Freud's work and advance the thesis of the visual constitution of the self. On visuality in French psychoanalysis see Chapter 6 in Jay, *Downcast Eyes*.

Chapter 4
"Seeing the World with the Eyes of God": An Alternative Explanation of "Reverse Perspective"

In his book *Art and Beauty in the Middle Ages* Umberto Eco says that: "they [the Medievals] saw the world with the eyes of God".[1] I would like to suggest one way in which this statement is true. I will be proposing in this chapter that some of the "distortions" which lie at the heart of "reverse perspective" are informed by a timeless conception of God's eternity. To a God who transcends the temporal dimension, events of human history exist simultaneously, all at once. By implication, such a timeless being will not perceive objects successively in time but simultaneously. In this sense, divine vision is simultaneous and thus "view-pointless", i.e., things are not seen from a certain point of view but, potentially, from all possible viewpoints at once. As God, or more precisely his "eye",[2] is not subject to spatial location he is also ubiquitous. In other words, God transcends space as well and his spacelessness becomes a metaphor for his timelessness (even though it does not literally illustrate timelessness). Space and time in this pre-Einstein context are not to be conflated automatically. They are distinct dimensions which are considered together through their shared transcendence in God's world.

On the other hand, an object perceived from a timeless and ubiquitous perspective is curiously reminiscent of the Platonic "Idea". According to Plato, a painted representation of an object is at a third remove from the Idea of the object, as it is a shadow of the shadow. A naturalistic representation of a bed – three legs, supporting an angular shape, jutting out in the direction of the viewer – is a "lie" to reality, i.e., this is how we may see a bed from a fixed point of view at a certain moment of time, but this is not what a bed *is* in absolute terms. Following the principles of iconic representation a bed could be represented both with its top, its underside on all four legs – it is true to reality and approximates the Idea, i.e., this may not be how we see a bed at fixed moment and position, but it is what a bed *is*. This is why, on the whole, icon art can be said to be resistant to the accidents of both space and time. Even when there are indications of location and/or moment these, in the great majority of cases, bear a generalized character.

---

[1] Umberto Eco, *Art and Beauty in the Middle Ages* (New Haven and London, 1986, first in Italian in 1959), p. 118.

[2] God's "eye" has been appropriately described by John of Damascus as the "divine, all seeing and immaterial eye" (for the quote see section B below).

To proceed with my argument, a re-definition of "reverse perspective" will be worked out by referring to a strand of thought in Florensky, which the Russian author left undeveloped. This view is distinct from the one prevalent so far and, rather ironically, largely promoted by Florensky himself. It is this definition, however, that brings to the fore the type of "distortions" we have been concerned with. Florensky's notion of the "supplementary planes" will be seen, on the one hand, in the context of ideas at the beginning of the twentieth century. More concretely, it will be suggested that it is developed on the background of experiments with pictorial space in Analytical Cubism, as well as notions of vision, usually associated with Theosophical writings.

While contemporary intellectual developments are of importance, what is probably even more significant is the possibility of analyzing the notion of what I will call "simultaneous planes" through the lens of theological notions which were part of the mental landscape at the time of the production of icons. The principle of organizing space in the icon will be considered within the framework of the doctrine of divine timeless eternity. A brief outline of some of the fundamental texts dealing with the concept of timelessness will be provided, starting from the classical Greek sources. It will be shown that in Antiquity there was an ambiguity on this issue, which led to swings between interpretations implying either the "everlasting" or the "timeless" views of eternity. The medieval understanding, however, can be traced almost directly back to pagan Neoplatonic philosophy and the "timeless" view promoted by writers as Plotinus. The ambiguities in the concept of eternity, however, persisted throughout the medieval period and to the present, where divine eternity is a consistent topic of debate addressed by some of the foremost philosophers of religion.

The present discussion focuses on the notion of timelessness which, however controversial in principle, was never decisively put in doubt by Eastern Orthodox theology. It also addresses the thorny problem of the relationship between the theological concept and the contemporary artistic practice of constructing images according to the principle of "simultaneous planes". Liturgy appears to be one of the obvious and most common means of transmitting concepts from "high" theology onto a level adapted to the wider audience of the faithful. There is a line of liturgical studies stressing the advent of eternity into time, which is realized particularly in the Eucharist. Another, more complicated, way of finding a common ground between theology and art is by looking at the possibility of a "structural intuition" that the two share. The recently re-sounded theme of "theology through the arts" implicitly accepts this possibility and thus lends support to our view that "reverse perspective" provides a structure in concrete visual terms which is analogous to the conceptual structure underlying timeless eternity.

I end by attempting to provide an intellectual context for my hypothesis. This includes tracing a line in post-medieval thought which witnesses the process of "a transformation of religious receptivity"[3] into an aesthetic one. We will be concerned with one aspect of this phenomenon, namely the development of the initially theological concept of simultaneity into a feature of aesthetic experience. Simultaneity becomes emblematic of a process, whereby art acquires metaphysical value through the notion of the aesthetic which transcends the spatial-temporal specificity of the image. The three selected authors are very much a matter of personal choice, based on what I felt was a common concern with the problem of a divine point of view. Leibniz (1646–1715) is important as marking the transition from the conception of timelessness as an exclusively divine characteristic to the possibility of its being an aspect of human experience. With Schopenhauer (1788–1860) art comes to the foreground in its role of transcending time, while with Wilhelm Worringer (1881–1965) it is shown how this happens specifically in ancient forms of non-naturalistic art.

A New Definition of "Reverse Perspective" as a Prerequisite for the Present Hypothesis

In Chapter 2 a critical reading of writings on "reverse perspective" was attempted, mainly in terms of discussing the underlying assumptions of the standard definition of "reverse perspective". In short, the conclusion was that "reverse perspective" cannot be accepted as turning around the laws of linear perspective in any of the implied senses. In this chapter I will endeavour to go one step further by asking the question: So what is "reverse perspective"? How can it be defined in more legitimate terms?

The predominant view, rejected in this paper, has been very much promoted by the Russian scholars we have discussed. At the same time, the hypothesis put forward in this chapter grows out of a certain aspect of this same tradition. My belief is that the phenomenon of "supplementary planes", pointed out first by Florensky and later by Zhegin and Uspensky, constitutes a genuine contribution to the theory of "reverse perspective". This observation was not exploited to the full by either author, but I think it is ground-breaking as it has the potential of leading us in the right direction. My aim in this chapter will be to elaborate on and develop further the implications, latent in the phenomenon of "supplementary planes". My starting ground will be to accept a definition of "reverse perspective" as *the simultaneous representation of different planes of the same image* on the picture surface, regardless of whether the corresponding planes in the represented objects could be seen from a single viewpoint. This definition will stand in contrast to the commonly accepted one ("reverse perspective" is, in some ways, an opposite of linear perspective). I will be using the term "simultaneous planes" henceforth, in

[3] Otto Pächt, 'The End of Image Theory' (1930–31), in Christopher Wood (ed.), *The Vienna School Reader: Politics and Art Historical Method in the 1930s* (New York, 2000), p. 191.

preference to Florensky's "supplementary planes", as I believe it to describe better the nature of the phenomenon under our attention.

As became clear both positions are present in the Russian writings on the subject – i.e., "reverse perspective" turns around the laws of linear perspective and "reverse perspective" is characterized by "simultaneous planes" – but they are never opposed, as the authors seemed unable to break from the standard view. Interestingly, no one was stuck by the idea that these two positions were incompatible.

It is of importance to notice that the standard definition of "reverse perspective" was uniformly accepted by scholars in the West, while no notice was paid and no serious mention was made of the phenomenon of "supplementary planes". There is no study devoted specifically to this problem outside Russia, but whenever referring to the phenomenon the same idea is monotonously repeated – "the vanishing point" in "reverse perspective" is "not behind but in front of the picture".[4] More rarely, Western authors think of "reverse perspective" as "hieratic", i.e., in the vein of Doehlemann. Such is the case with Miriam Bunim, who defines "reverse perspective" as a method of "hierarchic scaling", where "the figures are the smaller the nearer they are to the spectator rather than larger as in normal perspective vision".[5] *The Oxford Companion to Art* holds the same view and so "perspective in Byzantine art" is described as "hieratic and anti-illusionistic".[6] Again, it escapes attention that these two views are not at all the same and refer to completely different phenomena – in one case, it is claimed that there is a vanishing point in front the picture plane, in the other that the sizes of objects differ according to hieratic significance. In fact, the latter, as we saw, was proposed by Doehlemann as an alternative to the standard view, while most studies convey no awareness of the contradiction between the two explanations.

What is accepted in this book as a potentially fruitful and productive approach to the problem of "reverse perspective" was first intimated in the opening paragraphs of Florensky's essay 'Reverse Perspective'. Section I brings to attention the artistic fact that "the icon often shows parts and surfaces which cannot be seen simultaneously".[7] The examples that the author mentions have already been referred to in Chapter 2 – the adding up of the façade and two side views of

---

4 See Gervase Mathew, *Byzantine Aesthetics* (London, 1963), p. 33; Michel Quenot, *The Icon: Window on the Kingdom* (London, 1992, first in French in 1987), p. 106; Stuart Robinson, *Images of Byzantium* (London, 1996), p. 20; Egon Sendler, *The Icon: Image of the Invisible* (Redondo Beach, CA, 1988, first in French in 1981), p. 127, etc.

5 Miriam Bunim, *Space in Medieval Painting and the Forerunners of Perspective* (New York, 1940), p. 7.

6 Harold Osborne, *The Oxford Companion to Art* (Oxford, 1970), p. 856.

7 Pavel Florensky, 'Obratnaia perspektiva' (Reverse Perspective) in Pavel Florensky, *Khristianstva i kultura* (Christianity and Culture), ed. A.Filolenko (Moscow, 2001), p. 38; Pavel Florensky, 'Reverse Perspective' in Pavel Florensky, *Beyond Vision: Essays on the Perception of Art*, ed. N. Misler (London, 2002), p. 201.

buildings, the three or even four aspects of the Bible, the treatment of the face in which the front view is seen alongside the temple and the ears which appear "as if it were, spread out on the surface of the icon".[8] Florensky notices that the "additional surfaces" of the represented object, the ones that should not be there if following the laws of normal vision at a single moment in time, are frequently emphasized by means of colour. Not only are "additional"/"supplementary"/"simultaneous" surfaces not left in shadow but they are actually painted in strikingly bright colours that capture our attention.[9] The Russian author finds in this phenomenon of the simultaneous presentation of different aspects of the image on the same picture plane the explanation for "reverse perspective" and mainly the fact of the tendency of parallel lines to diverge: "In regard to these supplementary planes, lines that are parallel and do not lie on the plane of the icon, or lines that are parallel to it which should be shown converging on the horizon, are instead shown in an icon diverging from each other".[10] While, as mentioned in the previous chapter, I believe Florensky to have misrepresented the nature of the diverging lines in "reverse perspective" in the already discussed sections XIII–XIV of his essay, I see a genuine contribution in the significance the author bestows on the element of the "additional planes" in section I. In fact, when Florensky drew attention to the simultaneous representation of different aspects of the image on the same picture plane in icon art he did more than state an artistic fact. He was also redirecting the whole definition of "reverse perspective" by implying, at least at this initial stage of his essay, that this characteristic treatment of planes is the fundamental feature of the organization of the pictorial space in the icon.[11] As we saw, other Russian scholars followed in his step, even though they did not go beyond Florensky's observation.

Once an icon is seen in that light – as the spreading out of several sides of the image on the same picture plane – it is possible to ask what a conception of time is disclosed. That there is a sort of temporal dimension becomes obvious by the commonly made argument that "reverse perspective" is anti-naturalistic because objects just do not appear like that to human vision. It is noticed, with good reason, that it is impossible to see simultaneously different aspects of an object which can be seen only at different moments in time, one after the other. However, at the heart of a religious worldview lies the belief that what is impossible for man is quite possible for the omnipotent God. In particular, a timelessly eternal God to whom all moments in time exist simultaneously should be able to see all points in space simultaneously as well. If we accept the first part of the argument the second follows logically. Then it would be possible to maintain that an icon is constructed

8 Ibid., p. 38; Ibid., p. 201.

9 Ibid., p. 40; Ibid., p. 203.

10 Ibid., p. 39; Ibid., pp. 201–202.

11 Part of the problem with Florensky's essay is that later on in his text he reverts to the standard definition of "reverse perspective", which yields to a misinterpretation of the artistic phenomenon under discussion.

in a way that it could appear to divine vision or rather, the nearest that an image can come to conveying such a vision to our more limited sight.

It is worthwhile noticing that the notion of simultaneity was an important part of the cultural, scientific and artistic discourse of the *fin-de-siécle.*[12] More specifically, the principle of simultaneous planes of the icon reveals obvious structural analogies with the construction of space in Cubist paintings. Further, it has been noticed that this aspect of Cubist art has intellectual and structural links to notions of occult idealism. I would like to suggest that the occult concept of the astral plane in particular shows a striking resemblance to the simultaneous planes of the icon. That this new view of "reverse perspective" makes points of contact with Cubism and the revival of occult notions in the early twentieth century has not been remarked so far, mainly because it depends on an understanding of iconic space which, as we saw, has gone almost completely unnoticed. In other words, I am suggesting that Florensky's notion could be seen as contributing to an on-going dialogue at the time and was certainly shaped with a knowledge of this intellectual and artistic background.

What at the beginning of the twentieth century Henry Adams had called "the law of acceleration"[13] had become a characteristic of human experience since the Industrial Revolution. The sense of acceleration under the impact of the new technologies was such, however, that it frequently left the impression of something approaching simultaneity. Notions of time were drastically transformed as a result of the discovery of the telephone (1876), the cinema (1893), the wireless (1895), the X-rays (1895), etc.[14] This new, typically modern conception of time found a range of expressions in the arts. People like Blaise Cendrars (1887–1961) and Guillaume Apollinaire (1880–1918) were writing what they called "simultaneous poetry". James Joyce (1882–1941) was exploring various techniques in *Ulysses* to create effects of simultaneity. Music was the obvious model and composers like Igor Stravinsky (1882–1971) were experimenting with tri-tone harmonies to expand the impression of different melodies working together and at the same time. Cendrars's poem 'La Prose du Transsibérien et de la petite Jehanne de France', printed on a sheet two metres long was accompanied by Sonia Delaunay's (1885–1979) illustrations in *couleurs simultanées*.[15] The reader, who was also the viewer, was meant to read the poem, look at the illustrations, as well as at the added

---

[12] See Pär Bergman, *'Modernolatria' et 'Simultaneity': Recherches sur deux tendances dans l'avant-garde littéraire en Italie et en France á la veille de la première guerre mondiale* (Stockholm, 1962).

[13] Henry Adams, *The Education of Henry Adams: An Autobiography* (Boston, 1918, printed privately in 1907).

[14] See Stephen Kern, *The Culture of Time and Space 1880–1918* (London, 1983), pp. 65–89; Reinhart Koselleck, *Zeitschichten. Studien zur Historik* (Frankfurt, 2000).

[15] See Antoine Sidoti, *La Prose du Transsibérien et de la Petite Jehanne de France: Blaise Cendrars, Sonia Delaunay, nov.-dec.1912 –juin 1914; genése et dossier d'un polemique* (Paris, 1987).

map of the trip that Cendrars's poem celebrates all at the same time. Apollinaire's character Baron d'Ormesan could be simultaneously at different places and he ends by dying at 820 places at the same time.

The Cubist Background

The most important development in painting in particular, however, was Cubism and it is by far the most relevant to our discussion. It might appear too far-fetched to compare Picasso's *Les Demoiselles d'Avignon* (1907), "the painting that brought art into the twentieth century",[16] with medieval images. At the same time, the development of Russian Cubofuturism makes such a comparison highly likely. Painters like Natalya Goncharova (1882–1962) were increasingly concerned with the creation of a national school of modern art. This project involved a conscious distancing from European artistic influences and an identification with the age-old art of the icon, as well as with forms of Russian folk-art. In practice, a synthesis between foreign modernist influences and local art forms was worked out,[17] which in Goncharova's case, for example, was highly successful.[18] This development had a two-fold impact. On the one hand, it constituted an original contribution and re-interpretation of European Cubism. On the other, it furthered the rediscovery of the icon, taking place in Russia since the middle and the end of the nineteenth century.

My immediate purpose in this text is to draw attention to a fundamental structural similarity in the construction of pictorial space in Cubist painting (especially early Analytical Cubism) and in icons, which partly made possible the synthesis mentioned above. I suggest that there is a profound analogy behind the principle of "multiple planes"[19] of Cubist paintings and the "simultaneous planes" of Eastern Orthodox images. According to Arthur Miller, the former comes down to "the simultaneous representation of entirely different viewpoints, the sum total of which constitutes the object".[20] This definition sounds close to Florensky's idea and indeed it could well describe the construction of iconic images, according to the view proposed here. The squatting female figure in Picasso's painting reveals the superimposition of frontal and profile views on the same plane which is fundamentally similar to the same phenomenon that occurs in medieval images. Another work by Picasso, *Plate and Bowl* (fig.4.12 in Miller, p.115) brings naturally to mind – at least, to anyone used to viewing icons – the representation of objects in icon art. The plate is seen from above and from this point it appears circular,

[16] Arthur Miller, *Einstein, Picasso: Space, Time and the Beauty That Causes Havoc* (New York, 2001), p. 2.

[17] Camilla Gray, *The Russian Experiment in Art: 1863–1986* (London, 1986), p. 97.

[18] See her much admired Byzantine-inspired designs for Diagliev's *Ballet Russes* in Paris, especially the one for Rimsky-Korsakov's *Le coq d'or* in 1914.

[19] Miller, *Einstein, Picasso: Space, Time and the Beauty That Causes Havoc*, p. 106.

[20] Ibid.

while the bowl is seen from a side view. Other Cubist painters, as Jean Metzinger, continued playing on this principle, which went back to Cezanne (1839–1906). We already saw that the changing viewpoints in regard to objects or parts of objects within the same painting were a typical feature of the icon. The "distortions" in both cases – in Cubism and in icons – are largely due to this strategy. It is even more interesting to see that there is a common motivation – to represent as if the Platonic Idea of an object rather than an individual item. Cubists would call this, very appropriately, a "type". André Lhote, who was part of Picasso's circle in Paris, says:

> "You represent by means of a trapezoid a table, just as you see it, distorted by perspective, but what would happen if you decide to express the table as a type [la table type]? You would have to straighten it up onto the picture plane, and from trapezoid return to a true rectangle. If that table is covered with objects equally distorted by perspective, the same straightening up process would have to take place with each of them. Thus the oval of a glass would become a perfect circle."[21]

In other words, by painting the "type" of an object the painter abstracts from the various appearances of tables, glasses, etc. a single version which is taken to comprise the essential features of each object. How does the artist proceed about extracting a "type"? Pretty much like the medieval iconographer, he/she moves around "an object to seize it from successive appearances which, fused into a single image, reconstitute it in time".[22]

There is, however, one major difference between the Cubist painter and the medieval iconographer. Unlike the latter, Cubist artists were consciously in revolt against the Western pictorial tradition, epitomized by linear perspective. Braque was expressing a common attitude, shared by other Cubists when he said:

> "Traditional perspective did not satisfy me. Mechanical as it is, this perspective never gave the full possession of things. It started from a point of view and didn't leave it. But the point of view is one totally small thing. It is as if someone who all his life would draw profiles would come to believe that man has only one eye."[23]

In this sense, Cubism and icon art share what has been accepted as one of Cubism's major features – "the combination of several views of an object within

---

[21] Quoted in *op. cit.*, pp. 114–15.

[22] Albert Gleize and Jean Metzinger, *Du Cubisme* (Paris, 1912), p. 68; quoted in Miller, *Einstein, Picasso: Space, Time and the Beauty That Causes Havoc*, p. 169.

[23] Quoted in Miller, *Einstein, Picasso: Space, Time and the Beauty That Causes Havoc*, p. 130.

a single image".[24] However, the other characteristic of Cubist art – "the dismissal of a system of perspective which had conditioned Western painting since the Renaissance"[25] – could not have been a concern of any pre-Renaissance painter.[26]

In their reaction against the Western artistic tradition, the Cubist painters were conscious of themselves as innovators and even rebels. Florensky's text, on the other hand, could be interpreted as claiming that if one wants to get a "full possession of things" one should go back to the icon. Thus, the modernist project could be seen not only as a matter of innovation but also as a *re*discovery. In a way, Florensky's passages on the "supplementary planes" of the icon can be read as a hidden reference to Cubism and a claim that ancient Russian art had already achieved some of the principles that the avant-garde was experimenting with. Florensky was, of course, well aware of developments in modern art and the whole discourse on Cubism in particular at the time. Zhegin's memoir bears witness to Florensky's lively interest in modern art and his frequenting the meetings at the house of the Cubist painter L. V. Popov, where he would have met artists like Vladimir Tatlin.[27] Tatlin, a great admirer of Cubism, had actually met Picasso in 1913 and returned to Russia full of enthusiasm for the works he had seen at Picasso's studio. Apart from this, more than forty works by Picasso and by other French Cubists could be seen in the private gallery of the merchant Sergei Shchukin in Moscow at the time.[28] Florensky himself was well familiar with the Cubist paintings in Shchukin's Collection and actually wrote on Picasso's musical instruments there, which "represented objects from several points of view", so that we could appreciate these objects "in full, more profoundly and in a fundamentally new fashion".[29]

The French Cubists themselves may not have referred to medieval icons but they acknowledged the influence of various forms of "primitive" art on their work. Maurice Raynal comments that: "Instead of painting objects as they [the primitives] saw them, they painted them as they thought them, and it is precisely this law that the cubists have readopted, amplified and codified under the name 'the Fourth Dimension".[30]

---

24 John Golding, *Cubism. A History and an Analysis* (London, 1968), p. 27.

25 Ibid.

26 Hence, the misconception, pointed out several times in this book, underlying the term "reverse perspective", which implies a reversal of the rules of standard perspective.

27 Lev Zhegin, 'Vozpomenaniia o P. A. Florenskom' (Memories of P. A. Florensky) in Konstantin Isupov (ed.), *P .A. Florensky: Pro et Contra* (St. Petersburg, 1996), pp. 162–73.

28 Douglas Cooper, *The Cubist Epoch* (New York, 1971). The section on the influence of Cubism in Russia is on pp. 156–64.

29 Section X of *Smisl' idealizma* (The Meaning of Idealism) (1914), especially p. 102.

30 Quoted in Edward Fry, *Cubism* (London, 1966), pp. 129–30.

The Theosophical Background

It has been suggested that the notion of the "Fourth Dimension", which has become from early on so central in discussions on Cubism and other modernist movements, is to be understood in the light of occultism rather than in its strictly scientific sense, relating to Relativity Theory. This view was first suggested by Linda Henderson and John Adkins Richardson in the 1970s.[31] The same ideas that provided the intellectual background for the penetration of occult notions in Cubist circles in France could be traced in Russia at the time. A great deal of attention has been paid to the impact of occultism on Vasilii Kandinsky (1866–1944)[32] and Kazimir Malevich (1878–1935).[33] Florensky, who for a while worked with Kandinsky in the same department at the Russian Academy for Artistic Sciences (Malevich joined the same institution in 1924, the year that Florensky left), was exposed to similar intellectual influences which, in fact, were not confined to academic circles. It is, therefore, reasonable to propose that Florensky's understanding of the "simultaneous planes" of the icon was worked out with a knowledge of occult theories. More particularly, the notion of the "simultaneous planes" could serve, indeed could have been envisioned by Florensky, as a means of describing "astral vision" on the "astral plane", a conception associated particularly with writings in the Theosophical tradition.

The connection between Theosophy and Russian theory of the icon has not been made so far, but it is well known that the Theosophical movement had a decisive intellectual influence in *fin-de-siécle* culture in Europe in general and in Russia in particular. For a time, Theosophy became "the dominant alternative culture" and "the 'school' towards which artists and seekers could look for a radically *other* description of man".[34] The Theosophical Society itself and its founder Helena Blavatsky had an important impact on the work of Yeats, James Joyce, T.S. Eliot, and a host of other writers, poets, and artists at the time.[35] The pervasive influence of Theosophy in the case of Russia has attracted some systematic scholarly attention

---

31 Linda Henderson, 'A New Facet of Cubism: "The Fourth Dimension" and "Non-Euclidean Geometry" Reinterpreted', *The Art Quarterly*, 34 (1971): 410–33 and John A. Richardson, *Modern Art and Scientific Thought* (Urbana, 1971), Chapter 5: "Cubism and Logic".

32 See Sixten Ringbom,'Art in the "Epoch of the Great Spiritual". Occult Elements in the Early Theory of Abstract Painting', *Journal of the Warburg and Courtauld Institutes*, 29 (1966): 386–418. Also, by the same author, *The Sounding Cosmos. A Study in the Spiritualism of Kandinsky and the Genesis of Abstract Painting* (Abo, Finland, 1970).

33 See Susan Compton, 'Malevich and the Fourth Dimension', *Studio International*, vol. 187 (1974) and by the same author 'Malevich's Suprematism – the Higher Intuition', *Burlington Magazine*, CXVIII (1976): 577–85.

34 Roger Lipsey, *An Art of Our Own: The Spiritual in Art* (Boston, 1988), pp. 32–4.

35 Sylvia Cranston, *HPB: The Extraordinary Life and Influence of Helena Blavatsky, Founder of the Modern Theosophical Society* (New York, 1993).

only recently,[36] while the involvement of Symbolists like Viacheslav Ivanov and Andrey Bely with Theosophy and Spiritualism is well documented.

At first glance, Florensky is an unlikely candidate for an author to have been inspired by Theosophy. After all, he even took trouble to write personal letters to prominent members of the Theosophical Society in Russia, whom he denounced openly for their beliefs.[37] However, a close reading of his works reveals that, though not a Theosophist himself, Florensky had "read everything on the subject".[38] Like many other religiously-oriented thinkers at the time, Florensky could not accept the Theosophical Society in its guise of an anti-Christian philosophy which did not believe in a personal God, in the Biblical Creation, etc. At the same time, Theosophy as a movement of ideas which borrows from a variety of sources, such as Plato, Plotinus, Pseudo-Dionysius, but also more generally magical beliefs going back to Late Antiquity and even beyond, covers largely the material that Florensky himself was interested in. It is only in the last several years that the question of the relationship of Florensky's thought and occultism has been raised,[39] while Nicoletta Misler has suggested a link between Theosophy and the Russian author's notion of "reverse perspective". According to Misler, the first part of Florensky's essay represents an exploration of "transparent vision", a concept popularized by Theosophy.[40] None of Florensky's visual examples – or Zhegin's, which we looked at in Chapter 2 – actually confirm this view as no internal surfaces of objects are represented, as would have happened to a vision, which renders opaque surfaces transparent. Misler's point, though, is useful as it directs attention to the link between Florensky's view of iconic space and Theosophical notions of vision.

It has not been noticed so far that the essay on "Reverse Perspective" elaborates on ideas first explored by Florensky in an earlier work – *Smisl' idealizma* (The Meaning of Idealism) (1914). In *The Meaning of Idealism* and in the context of the discussion of Picasso's paintings of musical instruments in the Shchukin Collection

---

36 The most important studies are: Maria Carlson, *'No Religion Higher Than Truth': A History of the Theosophical Movement in Russia, 1875–1922* (Princeton, 1993) and Bernice Rosenthal (ed.), *The Occult in Russian and Soviet Culture* (Ithaca and London, 1997).

37 The draft of a letter (it is not known if the letter was sent) to Elena Pisareva has been preserved. In it, Florensky says: "I openly acknowledge to you that I believe all Theosophical and other journals, societies, brochures, etc. are a deeply negative and dangerous phenomenon" (*Florenskii i simvolisti* (Florensky and the Symbolists), compiled by Elena V. Ivanova (Moscow, 2004), p. 530; my translation).

38 Bernice Rosenthal, 'Introduction' in Bernice Rosenthal (ed.), *The Occult in Russian and Soviet Culture* (Ithaca and London, 1997), p. 20.

39 L. I. Vasilenko, L.I., 'O magii i okkul"tizme v nasledii o. Pavla Florenskogo'("On Magic and Occultism in the Heritage of Father Pavel Florensky"), *Vestnik Pravoslavnogo Sviato-Tikhonovskogo Gumanitarnogo universiteta*, 3 (2004): 81–99.

40 Nicoletta Misler, 'Pavel Florensky as an Art Historian', in Pavel Florensky, *Beyond Vision: Essays on the Perception of Art* (London, 2002), pp. 84–5.

in Moscow,[41] the author develops his notion of "synthetic vision". This notion is explicitly modelled on the background of the work of Charles Hinton (1853–1907), a writer usually discussed in the context of Theosophy, and it sounds remarkably close to the Florensky's own and later concept of "simultaneous planes". Consider the following passage: "The reality of the artistic image is realized in [...] unifying in one apperception that which is given at different moments and, consequently, under different angles of vision".[42] Thus, the essay on "Reverse Perspective" could be understood as putting up the icon as a model of "synthetic visuality". Therefore, Theosophical notions of visuality are considered important as one of the sources of Florensky's theory of the medieval image.

In 1895 the English Theosophist Charles Leadbeater (1847–1934) published a book, entitled *The Astral Plane: Its Scenery, Inhabitants and Phenomena*, in which he elaborated on ideas by earlier writers from the tradition of occultism. The most significant influence on Leadbeater was Charles Hinton, who had published several works[43] revolving around the problem of the possibility of achieving a vision of the transcendental and infinite. Hinton's works achieved great popularity at the time, including in Russia,[44] especially thanks to Petr Ouspensky's *Tertium Organum* (1911), whom Florensky cites. What Hinton called the "four-dimensional vision" corresponded to Leadbeater's "astral sight", which referred to "a faculty very different from and more extended than physical vision".[45] "Astral sight" allowed the clairvoyant to see an object "as it were, from all sides at once, inside of a solid being as plainly open to view as outside".[46] In the following text it will become clear that the expression "all at once" has a long history of describing the concept of simultaneity. Here, "astral vision" reveals what we called "simultaneous planes". The crucial characteristic of this kind of perception apparently did not depend on a point of view. Therefore, the sides of objects are flattened out in a manner that closely approximates the "simultaneous planes" of the icon. In fact, Leadbeater's famous illustration with the wooden cube carries the same meaning as Florensky's egg-shell which, in order to be represented on a two-dimensional surface, has to be broken and spread out, i.e., the convex form of the shell is flattened in the process. In his book *Clairvoyance* (1899) Leadbeater claimed that if you looked

---

41 On the Shchukin Collection, see Gray, *The Great Experiment: Russian Art 1863–1922*, pp. 63–5.

42 Pavel Florensky, 'Smisl' idealizma' (The Meaning of Idealism) in his *Sochineniia. v chetirekh tomakh* (Works in Four Volumes) (Moscow, 1999), p. 98.

43 The earliest being Hinton's article 'What is the Fourth Dimension?', *Dublin University Magazine* (1880): 15–34.

44 Tom Gibbons, 'Cubism and the 'Fourth Dimension' in the Context of the Late Nineteenth Century and Early Twentieth Century Revival of Occult Idealism', *Journal of the Warburg and Courtauld Institutes*, 44 (1981): 136.

45 Charles Leadbeater, *The Astral Plane: Its Scenery, Inhabitants and Phenomena* (London, 1895), pp. 3–4.

46 Ibid.

at a wooden cube with astral sight "you would see all the sides at once, and all the right way up, as though the whole cube had been flattened out before you, and you would see every particle of the inside as well – not *through* the others, but flattened out".[47]

The difference between physical sight and astral sight corresponds to the difference between physical plane and astral plane. On the physical plane objects are as if seen, more or less, according to linear perspective, i.e., "we see the further side [of a glass cube] in perspective – that is, it appears smaller than the nearer sidewhich is, of course, a mere illusion".[48] On the astral plane, however, the sides of the cube "would appear equal, as they really are".[49] Thus, Leadbeater concludes that astral vision "approximates much more closely to true perception than does physical sight",[50] which reminds us of Florensky's and later Russian authors' claim that "reverse perspective" is truer to vision – a claim that is never fully explained and is ultimately unconvincing. Much more importantly, however, Leadbeater implies that the "astral plane" gives a view of transcendental reality. It is this idea that has exercised a decisive impact on artists like Kandinsky and Mondrian (1872–1944).[51] In my view, Florensky could be understood to have interpreted the iconic image with its "simultaneous planes" as a variant of the astral plane. Thus, he was putting forward the icon and the vision it implied as a model for what Theosophy believed would be the vision of a future, nobler humanity.[52] Again, the thrust to the future, typical of *fin-de-siécle* modernism, was for Florensky also a bridge to the past, embodied in the sacred Russian image.

The whole discussion of "synthetic vision" in Florensky's book on idealism – specifically Platonic idealism – is taken up with the possibility of developing a "new habit of seeing" (Plato, *Republic*, 517 E), a problem that, as we saw, interested the Theosophists, too. Florensky sees the question already posed in Plato, most famously in the myth of the cave as: what does it mean "to see the ideas"? His reply comes down to the definition of the image as four-dimensional – the terminology sounds inescapably close to Theosophy – and as such possessing "a higher degree of reality".[53] This thinking, according to the Russian author, lies at the basis of a "generic method of looking at the world" which is interested in the

---

47 Charles Leadbeater, *Clairvoyance* (London, 1903, first in 1899), p. 37.

48 Leadbeater, *The Astral Plane: Its Scenery, Inhabitants and Phenomena*, p. 9.

49 Ibid.

50 Ibid.

51 See Ringbom, 'Art in the 'Epoch of the Great Spiritual'. Also, by the same author, *The Sounding Cosmos*.

52 A useful study on Theosophy, particularly in a Russian context, is Maria Carlson's *'No Religion Higher Than Truth'*. See also Andrei Rogozhin, *Put' teosofii* (The Path of Theosophy) (Petrozavodsk, 1992).

53 Pavel Florensky, *Smisl' idealizma* (The Meaning of Idealism) in his *Sochineniia v chetirekh tomakh* (Works in Four Volumes), vol.3 (Moscow, 1999), p. 108.

phenomenon "as a whole" and not only in "one moment of its history".[54] Modern man has lost exactly this ability to experience "the world as a unified being"[55] and, in this way, if art has a mission –and Florensky is at one with the German romantics in believing that it does[56] – it consists in restoring the ability to "see the wood behind the trees".[57] This is the meaning of "synthetic vision" and the principle of "supplementary planes" gives a visual expression of it.

The question of the extent to which the Cubists and occult scientists were familiar with the scientific dimension of the notion of the "Fourth Dimension" has attracted considerable scholarly attention.[58] It seems obvious that especially before 1911, i.e., before Relativity Theory became known, the "Fourth Dimension" could not have implied Eisntein's "four-dimensional space-time continuum".[59] This question does not seem all that relevant in our context, though. The simultaneity, implied in art – both in Cubism and in Byzantine and Byzantining iconography – seems to be of a kind that would not stand up to scientific analysis. As Einstein explained, absolute simultaneity was impossible in a universe with moving parts.[60] However, the notions of Cubist "multiple planes", the occultist "astral plane", and the "simultaneous planes" of the icon presuppose exactly absolute simultaneity. The implication is that, whatever the scientific position, the vision of transcendental essences was available to the artistic genius, to the clairvoyant and, in Florensky's view, to the medieval iconographer. The connection could be traced in Apollinaire's contention that "the past, present and future must be encompassed in a single glance" by the painter, by which act he/she becomes "aware of his own

---

54 Ibid., p. 110.

55 Ibid., p. 108.

56 I have discussed this in my article '"Seeing the World with the Eyes of God": The Revival of Romantic Theories of the Symbol in Pavel Florensky's Works', *Slavonica*, 14/1 (2008): 44–6

57 Florensky, *Smisl' idealizma* (The Meaning of Idealism), p. 115.

58 According to Paul Laporte, we can speak of parallel developments in art and science at the time, but "it would be completely misleading to assume that change in the pictorial idiom was the result of purely theoretical deliberation" (Paul Laporte, 'Cubism and Science', *The Journal of Aesthetics and Art Criticism*, 7/3 (1949): 253–4). Miller makes a strong case, on the other hand, for the Cubists' familiarity with developments in science (Miller, *Einstein, Picasso: Space, Time and the Beauty That Causes Havoc*).

59 Albert Einstein, *The Meaning of Relativity* (Princeton, 1923), p. 33.

60 Albert Einstein, 'On the Electrodynamics of Moving Bodies' (1905) in Hendrik Lorentz et al., *The Principle of Relativity: A Collection of Original Memoirs on the Special and General Theory of Relativity* (New York, 1952, first in English translation in 1923), pp. 42–3: "two events, which, viewed from a system of coordinates, are simultaneous, can no longer be looked upon as simultaneous when envisaged from a system which is in motion relatively to that system".

divinity".[61] In other words, modern artistic vision follows its own laws, formed at least partly on what had been hitherto associated with divine vision.

As we saw in Chapter 1, Florensky's approach to the problem of time is better understood when put in the context of the intellectual concerns of his time. Simultaneity, a concept growing out of theories on time and eternity, belongs to the same context. Similarly to the theory of "reverse perspective", the notion of simultaneity also relates to ideas, current in the early twentieth century. Unlike the former, however, simultaneity can be related to concepts in philosophy and theology from the era in which the images were produced.

Classical Greek Sources on Divine Eternity

The starting point for the discussion on time and eternity usually commences with Parmenides (c.510–450 BC), who has been attributed with the discovery of the notion of eternity.[62] Discussions of Parmenides' poem *The Way of Truth* could be said to provide the outline to the major definitions of eternity. The poem is not known to us in the original but through quotations in other authors. The version, translated by Richard Sorabji, is the following:

> "Nor was it ever, nor will it be, it now is
> all together, one, continuous."[63]

Sorabji draws out eight interpretations of these lines, but they basically deal with two contradictory views of eternity. "The everlasting interpretation" sees eternity as a duration, in which eternal things exist at all times. "The timeless interpretation" proposes that eternal things are timeless, they exist but at no times.[64] The two interpretations roughly correspond to what has been described above as the traditional/timeless and non-traditional/everlasting views on the relationship between time and eternity.

The timeless interpretation is subdivided by Padgett into a notion of "relative" timelessness and one of "absolute" timelessness. Thus, the author suggests that there are three options for understanding the meaning of Parmenides.[65] The first one is that "Being is unchanging and exists in time, forever" or, in other words, Sorabji's "everlasting interpretation". John Whittaker, for instance, subscribes to

---

61 Guillaume Apollinaire, *The Cubist Painters* (Harrogate, 2000, first in French in 1913), p. 20.

62 Richard Sorabji, *Time, Creation and the Continuum* (Ithaca and London, 1983), p. 98; Alan Padgett, *God, Eternity and the Nature of Time* (New York, 1992), p. 41.

63 Sorabji,*Time, Creation and the Continuum*, p. 99.

64 Ibid., p. 103.

65 Padgett, *God, Eternity and the Nature of Time*, pp. 39–40.

that option.[66] Padgett finds the second option most likely, however – "Being is unchanging and timeless, having its own unchanging duration". In other words, Padgett credits Parmenides with what he call a "relative" notion of eternity, according to which Being is beyond Measured Time,[67] which applies only to the world of Becoming. The third option would be that "Being is unchanging and timeless and non-durational". This is strictly speaking the traditional notion of eternity.

Gerald Whitrow's presentation of Parmenides' theory on time makes it clear why Parmenides has been seen by many[68] to have influenced Plato on that problem. The world of appearance, which is subject to time and to change, is contrasted with the world of becoming, which does not change and is timeless. We receive information about the former through our senses, which are deceptive, while the later is revealed to us only by reason, which is the only true mode of knowledge.[69]

The question of Parmenides' influence on Plato (*c*.427–348 BC) is very much a matter of what Parmenides' theory of time and eternity is meant to imply and what Plato's. Just as in case of Parmenides there is no unanimously accepted opinion about Plato's view on time and eternity. On the one hand, Plato seems to say that eternity is a category which has nothing at all to do with time, as in: "These all (months, days, and years; past or future existence) apply to becoming in time, and have no meaning in relation to eternal nature, which ever is and never was or will be" (*Timaeus*, 37E). On the other hand, the well-known "time is the moving image of eternity" (*Timaeus*, 37D) allows of more than one interpretation, among which the one that implies duration and everlastingness makes sense. Whether we should see time as an imitation of eternity, falling short of it but still sharing the common characteristic of duration or as no more than a caricature of eternity, having nothing to do with it in a fundamental sense is a matter of choice, dictated usually by the interpretation of a larger context. Harry Wolfson believes the relationship

---

66 John Whittaker, *God, Being, Time: Two Studies in the Transcendental Tradition in Greek Philosophy* (Oslo, 1971), pp. 16–22.

67 The concept of Measured Time is proposed by Padgett as a compromise view (to mediate between the everlasting and timeless interpretations). It refers to metric, clock time. God's topological time, on the other hand, implies that there is no fundamental distinction between finite and infinite time. Finite time, according to that understanding, would describe finite periods of equal length (e.g. half an hour), while infinite time refers to infinite periods of unequal length (e.g. before and after). The latter would be applicable to God, while the former would not. For a more detailed exposition of the concept see Padgett, *God, Eternity and the Nature of Time*.

68 Padgett, *God, Eternity and the Nature of Time*, p. 41; A. H. Coxon, *Philosophy of Forms: An Analytical and Historical Commentary on Plato's 'Parmenides' with a New English Translation* (Assen, Netherlands, 1999), p. 26: "The theory [of Forms] was a pluralistic development of Parmenides' monism".

69 Gerald J. Whitrow, *Time in History: Views on Time from Prehistory to the Present Day* (Oxford and New York, 1989), p. 40.

between time and eternity in Plato is one of antithesis, while the notion of eternity as everlastingness we encounter in Aristotle (for example, the *Physics*, 4.12–14).[70] Tamar Rudavsky interprets the *Timaeus* in the same fashion and points out the two words used by Plato – *aionios* which refers to eternity and *aidios*, which is properly translated as everlastingness.[71] On that Rudavsky is citing G. R. Lloyd's article 'Views on Time in Greek Thought', which draws attention to the terminology to defend the author's proposition that for Plato the sphere of time is antithetical to the sphere of eternity.[72] In the same vein, Leonardo Taran calls Plato "the first atemporalist",[73] while Sorabji draws attention to passages in the *Parmenides* (141 A5, A6, C8, D5) that explicitly describe the One as existing beyond time.[74] At the same time, however, Sorabji finds the opposing view on eternity possible as well. The interpretation of Plato's notion of eternity as everlastingness is supported, for instance, by Francis Cornford in *Plato's Cosmology*,[75] by W. von Leyden in his article 'Time, Number and Eternity in Plato and Aristotle'[76] and by Arthur Armstrong's 'Eternity, Life and Movement in Plotinus' Accounts of *nous*'.[77] The impressive number of adherents to both sides of the argument about Plato's notion of time and eternity gives credence to Sorabji's statement that "Plato allows implications of timelessness and of duration to stand side by side in his account of eternity without offering a solution".[78]

The one who could be credited with resolving the ambiguities left by Plato is Plotinus (*c*. AD 205–270), though even he is not always consistent.[79] On the whole, however, Plotinus was the first who consistently attempted to clear the ground and, in this way, drew the guiding lines for future discussion. The *Enneads* 3.7 is specifically devoted to the problem of time and eternity. According to Plotinus, the

---

[70] Harry Wolfson, *The Philosophy of Spinoza* (New York, 1969). See also Sorabji on that (Sorabji, *Time, Creation and the Continuum*, pp. 125–7) and Leftow (Brian Leftow, *Time and Eternity* (Ithaca and London, 1991), p. 1).

[71] Tamar Rudavsky, *Time Matters: Time, Creation and Cosmology in Medieval Jewish Philosophy* (Albany, 2000), p. 10.

[72] G. R. Lloyd, 'Views on Time in Greek Thought' in Louis Gardet et al. (ed.), *Cultures and Time* (Paris, 1976), p. 138.

[73] Leonardo Tarán, *Parmenides: A Text with Translation, Commentary, and Critical Essays* (Princeton, 1965), p. 175.

[74] Sorabji, *Time, Creation and the Continuum*, p. 108.

[75] Cornford, F., *Plato's Cosmology: the Timaeus of Plato* (London, 1937), pp. 98 and 102.

[76] Leyden von, W., 'Time, Number and Eternity in Plato and Aristotle', *Philosophical Quarterly*, 14 (1964): 35–52.

[77] Arthur Armstrong, 'Eternity, Life and Movement' in *Le Néoplatonisme* (Paris, 1971), pp. 67–74.

[78] Sorabji, *Time, Creation and the Continuum*, p. 111.

[79] Ibid., p. 114. Also Arthur Armstrong (ed.), *The Cambridge History of Later Greek and Early Medieval Philosophy* (Cambridge, 1967), pp. 246–7 and p. 455?

essence of time is not motion, as Aristotle taught (Aristotle, *Physics* 4: time is the measure of motion). Rather, the essential characteristic of time is duration – "the life of the Soul is a motion of change from one stage of life to another" (*Enneads*, 3.7.11). Eternity, on other hand, is totally devoid of duration, it "does not get its being from any extent of time, but is "prior" to any extent of time" (*Enneads*, 3.7.6).[80] In short, time and eternity are opposed as something which is extended versus something which cannot be extended (*Enneads*, 3.7.2). The former belongs to the realm of becoming and the latter to the realm of being. The relationship between time and eternity is analogous to the relationship between Becoming and Being, Illusion and Reality, Lie and Truth. As Plotinus says, "you must not then join what has being to what does not, nor time or temporal existence to eternity, nor must you stretch out the unextended" (*Enneads*, 1.5.7 (20–31). As eternity lacks any kind of duration, it can be defined as "a life that abides in the same, and always has the all present to it, not now this, and then again that, but all things at once [...] it is something which abides in the same, in itself and does not change at all but is always in the present" (*Enneads*, 3.7.3). And further, again following from the lack of temporal duration, eternity has "neither any earlier nor any later about it, but 'is' is the truest thing about is" (*Enneads*, 3.7.6 (23–36). In other words, Plotinus developed the extremely influential for Christian theology notion of an eternal Now that coexists with all earthly nows. It is that idea that forms the core of the traditional notion on eternity that from Plotinus and via Augustine and Boethius left its stamp on the entire course of Christian thought.

The logical question that arises once we place time and eternity in such an antithetical relationship is how we can describe the nature of eternity, as the use of temporal terms seems illegitimate in that context. Plotinus was well aware of that difficulty and he tackled it in a brilliant way – by giving a non-temporal sense to words which in their original context denote temporal notions. In *Enneads*, 3.7.6 (23–36), quoted above he uses this device in respect to "is", which he explains is being used "in the sense that it is by its essence and life". As Sorabji notices, "always" receives a similarly non-temporal connotation.[81] As the author mentions, it is interesting that almost at the same time and independently, Origen employs "always" in exactly the same fashion (*On First Principles*, 1.3.4).[82] The method of giving a non-temporal sense to a word was already known through Plato and Aristotle. In the *Timaeus*, Plato speaks of the soul as "prior" and "older" than the body not only in birth but in excellence (i.e., quality) as well. Aristotle does the same in respect to "prior" and "posterior" on many occasions.[83]

---

80 The inverted commas are mine and indicate that "prior" here could not be taken in its literal sense.

81 Sorabji, *Time, Creation and the Continuum*, p. 112.

82 Ibid.

83 See the references to particular works by Aristotle in Sorabji, *Time, Creation and the Continuum*, p. 114. See later authors who use the same device in ibid., pp. 115–17.

Christian Sources on Divine Eternity – the Timeless View

As Richard Swinburne has pointed out, "the doctrine of God's timelessness seems to have entered Christian theology from Neoplatonism, and there from Augustine to Aquinas it reigned".[84] However, the idea that the notion of timeless eternity is a Greek invention combined with the lack of Scriptural evidence for timelessness has proved uncomfortable for upholders of the traditional view. Scholars, such as Nelson Pike, have been quick to notice that "Plato (probably) thought that things of ultimate value are eternal in the sense of timeless. But Plato was not a Christian – nor can I think of any reason why a Christian should accept Plato's judgement on this matter without careful consideration of how it relates to the broad Christian tradition concerning the nature of God".[85] At the same time, from his in-depth analysis of the terminology, relating to time in the Bible, Oscar Cullmann has concluded that in primitive Christianity time is not a notion opposed to God, but rather God uses time as a means to reveal himself.[86] On the other hand, while agreeing that a timeless God was in no way discussed in the Bible, Helm finds "nothing unbiblical about the idea of a timeless God" and "no reason why a timeless God should not be an element in Christian theology".[87] He believes that while "the Bible does not teach it [...] [it] teaches many other things that make the idea of timelessness a reasonable theological concept to employ".[88]

Be that as it may, the still raging debate of the timeless versus everlasting interpretation of eternity is beyond our immediate concerns. What matters is that in the medieval period, both East and West, the reigning doctrine was that of timelessness.

For reasons of clarity I will subdivide the following section into a part, dealing with the "Latin tradition" and one, concerned with the "Greek tradition". There are no fundamental differences between the two when it comes to the problem under our attention to necessitate such a division. In fact, all the Greek and Byzantine theologians I cite were known to the West.[89] Issues of influence, however, have proved a complicated matter so I have preferred to make the mentioned division.

---

[84] Richard Swinburne, *The Coherence of Theism* (Oxford, 1977), p. 217. The same view has been upheld by William Kneale in his 'Time and Eternity in Theology', *The Proceedings of the Aristotelian Society*, 61 (1960–1961): 87–108 and Nicholas Wolterstorff in his 'God Everlasting', in C. Orlebeke and L. Smedes (eds), *God and the Good: Essays in Honour of Henry Stob* (Grand Rapids, 1975).

[85] Nelson Pike, *God and Timelessness* (London, 1970), pp. 189–90.

[86] Oscar Cullmann, *Christ and Time: The Primitive Christian Conception of Time and History* (London, 1957, first in 1951), pp. 37–50.

[87] Paul Helm, *Eternal God: A Study of God without Time* (Oxford, 1988), p. 2.

[88] Ibid., p. 11.

[89] See Gervase Mathew, 'Abstract of a Lecture on the Character of the Theological Divergence between East and West', *Eastern Churches Review*, 4–5 (1972–1973): 116–18; Stephen Runciman, 'The Schism between Eastern and Western Churches', *Anglican*

The Latin Tradition

Henri Bergson's distinction between metric time and the inner *durée* or the time of our consciousness is an idea already familiar through Augustine (354–430). As Umberto Eco notices, it is worth remembering that Augustine first explored that notion and "his pages on time remain amongst the most modern, precise and revealing on the subject in the entire philosophical tradition"[90]. Augustine's views on the relationship between time and eternity, however, are not that strikingly original and they are mostly directly borrowed from Neoplatonism. Augustine was well aware of this Neoplatonic influence on his thought and explicitly mentions it on a few occasions (*Confessions*, 8.2, *The City of God*, 8.5, 12; 9.10). In the context of our discussion it seems reasonable to agree with Padgett that there is "little in [Augustine's doctrine of eternity] which cannot find a parallel in Plotinus".[91] I will mention several major points made by Augustine, mainly in view of his huge influence on the subsequent Christian tradition to which he popularizes basically Neoplatonic ideas. My main reference will be Brian Leftow's chapter on Augustine in his book *Time and Eternity*.[92]

Let us follow two fundamental arguments that Augustine makes in support of God's timeless eternity, both of which carry Neoplatonic associations and were to become leitmotifs in the traditional view on eternity. The first is based on the claim that true existence is immutable existence, the second comes down to the doctrine of divine simplicity. As Augustine teaches,

> "Being is a name for immutability. For all things that are changed cease to be what they were, and begin to be what they were not. Nobody has true being, pure being, real being except the one who does not change."[93]

So, the first definition of God would be as an immutable entity. Moreover, "God [...] *truly* exists because He is unchangeable".[94] In other words, Augustine makes

Theological Review, XLIV/4 (1962): 337–50. On John of Damascus's influence in the West, see Adolf von Harnack, *Lehrbuck der Dogmengeschichte*, vol. 2: Die Entwicklung des kirchlichen Dogmas, 5th edition (Tubingen, 1951), p. 509 ff; Moshe Barasch, *Icon: Studies in the History of an Idea* (New York and London, 1992), p. 190. On St. John as a model for Aquinas see Karl Krumbacher, *Geschichte der byzantinischen Literatur* (Munich, 1891), pp. 174, p. 206.

90 Kristen Lippincott with Umberto Eco, Ernst Gombrich et al., *The Story of Time* (London, 1999), p. 12.

91 Padgett, *God, Eternity and the Nature of Time*, p. 43.

92 Leftow, *Time and Eternity* (Ithaca and London, 1991).

93 Augustine, Sermon 7.7 in Jacques-Paul Migne (ed.), *Patrologia Latina* (Paris, 1841), vol. 38, p. 65.

94 Augustine, 'The Nature of God', 19 in *S. Aur. Augustini: Opera Omnia* (Paris, 1836), vol. 8, p. 780, the italics are mine.

a typically Platonic point – there are degrees of existence in terms of truthfulness. Genuine and thus real (in the sense here) existence is the changeless and timeless one, which belongs exclusively to God. Temporal existence is subject to change and thus less real. Augustine explicitly compares the two modes of existence: "God so exists that compared with Him, things which were made do not exist. If they are not compared with Him, they exist, since they are from Him. But if they are compared, they do not exist, because to exist truly is to exist immutably, which He alone is".[95] We come across the immutability argument throughout the centuries and till now with the adherents to the traditional view. Thomas Aquinas states that: "being eternal follows from immutability [...] Hence since God is maximally immutable, it supremely belongs to Him to be eternal" (*Summa Theologica*, Ia 10, 2).[96] More recently, Paul Helm has reaffirmed the same view by saying that God's being immutable would necessarily entail God's being timeless.[97]

The claim that timeless beings are "more real" than temporal ones may be difficult to accept by a modern reader. Leftow, however, finds that "it makes sense and is surprisingly plausible".[98] The author's efforts to prove that statement are outside our immediate interests, but his understanding of the claim of Augustine and the Platonists deserves mention. That timeless entities possess a greater degree of reality in comparison to temporal ones could mean two things, both of which are possible. Firstly, a timeless being exists in a kind of present which knows no past or future. From that follows that it is more genuinely present than a temporal being would be, as its presence is only that – present without the other two modes of past and future.[99] Secondly, as a timeless being does not change and is always the same, its presence is the same at any moment of time, i.e., nothing gets lost in the case of a timeless being. At no time will it be anything different or present in a different way than it is. In that sense it is fully present in contrast to a temporal being, whose presence changes over time.[100]

So far God has been described as timeless, immutable and more real than his temporal creation. As he is the only being who is, by definition, beyond time and is therefore absolutely immutable and possesses the greatest degree of reality, all these terms can be used as synonyms to "God". Another one that should be added to the list in this context is that God and only God is absolutely simple, a definition which naturally follows from the others mentioned above. The second argument of Augustine is that a being that exists beyond space and time is not composed of

---

95 Augustine, 'Commentary on Psalm 134', 4 in *S. Aur. Augustini: Opera Omnia* (Paris, 1836), vol.4, p. 2130.

96 The term "eternal" here, as elsewhere with Aquinas, means "timeless". Aquinas never uses the term "timeless" itself and so brings further confusion to the debate. I thank Prof. Swinburne for this note.

97 Helm, *Eternal God: A Study of God without Time*, p. 90.

98 Leftow, *Time and Eternity*, p. 4.

99 Ibid., p. 84.

100 Ibid., p. 85.

scattered parts; it cannot come apart spatially and has its duration all at once, in an eternal present. Consequently, it could be said that a timeless being possesses a greater inner unity in comparison to temporal beings, which are composed of parts in space and have a duration in time. As God has such a unity in the highest degree he is also defined as absolutely simple.

What is probably considered as the classical definition of divine eternity is the one given by Boethius (c.480–525) in *The Consolation of Philosophy*, 5.6.[101] I will be using here the translation by Eleonore Stump and Norman Kretzmann. Eternity is then defined as that which

> "includes and possesses the whole fullness of illimitable life at once and is such that nothing future is absent from it and nothing past has flowed away [...] and of this it is necessary both that being in full possession of itself it be always present to itself and that it have the infinity of mobile time present (to it)".[102]

According to Alan Padgett with Boethius "we find a clear definition, an explicit distinction of terms, and the use of doctrine to solve a perplexing theological problem".[103] This, however, hardly seems to be the case, as the heated debates in contemporary scholarship on the exact meaning of this passage show. If there is anything problematic about that description it is mainly carried by two expressions from the same sentence that may sound ambiguous and even contradictory. It is possible to interpret "all at once" as meaning "in an instant", while "illimitable" to mean "enduring forever in time". This "all at once" can be taken to refer to timeless eternity while "illimitable" would be one of the descriptions of everlasting eternity. Stump and Kretzmann have suggested the passage implied some sort of duration,[104] a view against which Paul Fitzgerald reacted in an article.[105] Stump and Kretzmann have attempted to clarify their position and, in short, have interpreted Boethius' view on eternity as one involving *atemporal* duration or what they call, E-duration. In contrast to temporal duration, duration in eternity does not entail successiveness and divisibility. By putting forward this highly obscure notion, the authors hope to avoid the possible charges that duration would imply that God is not absolutely immutable and simple (Augustine's two arguments, mentioned above). The notion of E-duration, and particularly in connection with Boethius, has been felt by many to be unconvincing. According to Helm, Boethius' definition has

---

[101] See Gerald Louglin, 'Time' in Hastings, Mason and Pyper (eds), *The Oxford Companion to Christian Thought*, (Oxford and New York, 2000), p. 707.

[102] Eleonore Stump and Norman Kretzmann, 'Eternity', *Journal of Philosophy*, 78/8 (1981): p. 430.

[103] Padgett, *God, Eternity and the Nature of Time*, p. 46.

[104] Stump and Kretzmann, 'Eternity', 430.

[105] Paul Fitzgerald, 'Stump and Kretzmann on Time and Eternity', *Journal of Philosophy*, 82/5 (1985): 82.

"nothing to do with duration",[106] just as Sorabji finds the interpretation that implies duration to be wrong.[107] In the same way, Padgett sees Boethius as having given "the classic expression to the idea of an absolutely timeless divine eternity",[108] which is, for the author, as became clear, non-durational. Padgett refers us to another text by Boethius, *On the Trinity*, where a qualitative distinction is made between "sempiternal" life and eternal life, the former implying a notion of everlastingness in time, the latter – timelessness.[109] The term "sempiternal" Boethius actually borrows from Augustine with whom it means not full eternity and refers to angels and the souls of the dead, but not to God. Actually, if we accept that Plato was unclear on the problem of eternity, so was Boethius, who describes time, in a similar fashion, as the imitation of eternity. In *The Consolation of Philosophy*, 5.6, time is defined as "the infinite changing of things", which is "an attempt to imitate this state of presentness of a changeless life, but since it cannot equal this state it falls from changelessness into change, from the simplicity of presence into the infinite quantity of past and future".[110] Time here would be taken to mean the failed imitation of eternity – it attempts to be present in the way that eternity is but fails to remains so. At the same time it is reasonable to interpret eternity itself to mean being present "all at once", i.e., living the whole of its life all at once. Leftow, however, analyzes that passage and the previous from *The Consolation* in terms of duration. He does not seem to me to have made a very convincing case, but is worth mentioning within the context of the opposing views about the meaning of what is probably the most influential definition of the relationship of time and eternity in the Middle Ages.

The influence of Boethius is evident with almost every major medieval writer, discussing the problem of time and eternity. With Anselm (1033–1109), according to Padgett, "the influence of and similarity to Boethius' definition of eternity is obvious".[111] We can accept that view, however, only if we understand Boethius as having implied a timeless definition of eternity, as Anselm seems to have understood the problem in those terms. In the *Proslogion*, chapter 19, Anselm says of God:

> "None of your eternity is past as if it now did not exist, and none of it is future, as it did not exist yet. Therefore it is not the case that you existed yesterday or will exist tomorrow. Instead, yesterday, today, and tomorrow, you exist. Or better,

[106] Helm, *Eternal God: A Study of God without Time*, p. 40.

[107] Sorabji, *Time, Creation and the Continuum*, pp. 199–200.

[108] Padgett, *God, Eternity and the Nature of Time*, p. 44.

[109] Ibid., p. 46.

[110] Boethius, *The Consolation of Philosophy*, 5.6 in Steward and Rand (eds.), *Boethius*, 11.9–11, 40–53, pp. 400 and 402.

[111] Padgett, *God, Eternity and the Nature of Time*, p. 47.

> neither yesterday nor today nor tomorrow do you exist, but you simply exist, beyond all time. For yesterday, today and tomorrow exist only in time."[112]

God's timeless eternity with Anselm, as with other upholders of the traditional view, follows from his greatness – "For nothing contains You, but You contain all things"[113] and from his simplicity, since he is indivisible. Anselm, however, pays attention to another aspect of the relationship between the timeless and the temporal. While God is outside time, temporal entities, according to Anselm, exist not only in time but also in eternity. This, Leftow maintains, is the key that Anselm provides for the relation between time and eternity.[114]

Usually, Thomas Aquinas' (c.1225–1274) discussion of divine eternity is regarded as the culmination of the Augustinian-Boethian tradition.[115] In the *Summa Theologica*, Iq.10, Aguinas explicitly states his indebtedness to Boethius' definition. Basically, Aquinas' argument runs as follows – God is simple (*Summa Theologica*, Ia, q.9, a.1), consequently he is changeless, from which it could be concluded that he is timeless, since time is the measure of change. The idea that God is timeless means that he coexists with all the modes of time – past, present, future. Aquinas, however, suggests a way of coming out of the difficulty posed by the problem of the relationship of a timeless being to its temporal creation. There is a possibility, he teaches, of timeless actions which produce temporal effects, without changing the eternal cause (*Summa Theologica*, Iq.10). The suggestion is important as it addresses a genuine problem and gives a solution which was much elaborated on and revised by later adherents to the timeless view.

The Greek Tradition

In the Eastern part of the Empire, there seem to have been no serious divergences from the view as outlined so far. At the time of Augustine, Basil the Great (339–379) defines the problem of eternity in similar terms and this comes as no surprise in view of the common sources, namely Plotinus and Origen. The *Hexaemeron* is a sermon on Genesis 1:1:26. It opens with an account of the creation of the heavens and earth. The bodies in heaven move in a circular motion. Even though "the beginning of the circle is not easily discerned by our means of perception"[116] we know that there is such a beginning. The argument for that statement is as simple as: "even if it does escape our observation, assuredly, He who drew it [the

---

112 Anselm, *Proslogion*, chapter 19 in F. Schmitt, (ed.), *S. Anselmi: Opera omnia* (Edinburgh, 1946), Opera, vol.1, chapter 19, p. 115, 11.10–14.

113 Ibid.

114 See Leftow, *Time and Eternity*, Chapters 10 and 11.

115 See Loughlin, 'Time', p. 708.

116 Basil the Great, 'On the Hexaemeron' in *Fathers of the Church*, vol. 46: *St. Basil: Exegetic Homilies* (Washington, 1963), p. 6.

circle] with a centre and a radius truly began from some point".[117] If the motion has a beginning it would necessarily have an end. Thus the world is not eternal but rather exists in time and is subject to genesis and destruction. While the world and time have a beginning in the act of creation by God there was something before[118] this world, "a certain condition older that the birth of the world and proper to the supramundane powers, one beyond time", "without beginning or end".[119] The two "conditions" are thus contrasted – the temporal one of the earth and earthly things and the eternal of God. The former is defined by change and transience. In a passage, reminiscent of Augustine, the nature of time is defined as that "whose past has vanished, whose future is not yet at hand, and whose present escapes perception before it is known".[120] The eternal condition is characterized exactly by its timeless characteristics and these are, as with Augustine and others, immutability and simplicity. Not only God but the angels too do not presuppose time and so are immutable. The nature of time underlies the week, while the eighth day brings associations with eternity – "a day without evening, without succession, and without an end [...] because it lies outside this week of time".[121] There is no change in the eighth day, it is all "one" as the first day, which was created outside time. It is in that sense that Basil interprets the Biblical "one day" in "and there was evening and morning, one day", where "one" is "an image of eternity"[122] and "day" is synonymous with "age".[123] With both terms the author stresses the connotation of "the unique and not the manifold",[124] which is typical of a timeless eternity. This is how we come to the second argument for divine timelessness, that of divine simplicity, which runs through all of St. Basil's works and especially *On the Holy Spirit*. In his discussion of Basil's works Georges Florovsky produces a useful exposition of Basil's idea of divine simplicity as a characteristic of the timeless being of God. Division is the outcome of time, while beyond time and in God there is complete and absolute unity. Indivisible wholeness, Basil maintains, is a proof that the Trinity is divine.[125] Simplicity here, as with the mentioned writers of the Latin tradition is an aspect of greatness. The Spirit, according to Basil, is

---

117 *op. cit.*, p. 7.

118 Notice the implication of priority in time, which could cause confusion.

119 Basil the Great, 'On the Hexaemeron' in *Fathers of the Church*, vol. 46: *St.Basil: Exegetic Homilies* (Washington, 1963), p. 9. The English translation, which uses the term "everlasting" confuses the issue as Basil clearly means "timeless" here.

120 Ibid.

121 Ibid., p. 35.

122 Ibid.

123 Ibid.

124 Ibid.

125 George Florovsky, *Collected Works* (Belmont, Mass., 1972–1989), vol. 7: *The Eastern Fathers of the Fourth Century*, p. 105.

an "intellectual being, endless in His strength and boundless in His greatness. He cannot be measured by time or by all the ages".[126]

With Gregory Nazianzen (330–389) we have the same opposition between earthly time, which is subject to change, and divine eternity, which is completely beyond time. Florovsky's discussion of St. Gregory pays special attention to that aspect of the Greek theologian's teaching. On various occasions Gregory Nazianzen expresses the doctrine that "that which is temporal is not God".[127] The main argument of divine unity and simplicity is already familiar to us – God is eternal by nature as "He contains within Himself the whole of being which has no beginning and will never end".[128] In 'Oration 29' Gregory says that "things which produce Time are beyond time",[129] while 'Oration 30' on the nature of the Son, says that the Son, similarly to the Father "exists outside time and absolutely".[130] The main point of interest, however, in 'Oration 29' falls on the problem of terminology. In the vein of Plotinus, Gregory is concerned with the vocabulary we use when describing divine eternity. Precisely because God's existence is beyond time it is incorrect and misleading to apply terms to it that imply time. In the spirit of the Cappadocian school,[131] Gregory is aware of the limitation of human reason and the impossibility of human language to describe divine essence, which is ineffable ('Oration 28'). The author notices, like others before him, that "expressions like 'when', 'before x', 'after y' and 'from the beginning' are not free from temporal implications however much we try to wrest them[132]". Thus Gregory finds no way out of the problem, as "we cannot explain […] the meaning of 'supratemporal' and deliberately keep clear of any suggestion of time".[133] As a compromise, he suggests that we use the term "world-era" in reference to God's eternity, which would mean "the period coinciding with the eternal thing".[134] In this case, the word "period" should be stripped of its temporal nuance, as unlike "time" it is not "to be measured or fragmented by the Sun's motion".[135] Whether Gregory's tentative attempts at a revision of terminology are successful or not is not as important as the fact that he draws attention anew, as Plotinus had done earlier, to the difference in kind between time and eternity and the consequent

126 Basil the Great, 'On the Hexaemeron' in *Fathers of the Church*, vol. 46: *St.Basil: Exegetic Homilies* (Washington, 1963), p. 35.

127 Ibid., p. 129.

128 Ibid., p. 31.

129 Gregory Nazianzen, *Faith Gives Fullness to Reasoning: The Five Theological Orations of Gregory Nazianzen* (Leiden, New York, 1991), p. 247.

130 Ibid., pp. 268–269.

131 The Cappadocian school consists of Basil the Great, Gregory Nazianzen and Basil's brother, Gregory of Nyssa.

132 Gregory Nazianzen, *Faith Gives Fullness to Reasoning*, p. 246.

133 Ibid.

134 Ibid.

135 Ibid.

problem of defining the latter in human terms. When discussing the begetting of the Son the author maintains that it took place in "a non-temporal manner".[136] In this sense, the question "whether or not what has been begotten from eternity existed prior to its begetting" is misplaced as "that question only arises in connection with temporally determined beings".[137] As an initial approach to the problem, Gregory advocates "drop your idea of flux, division and cleavage, drop the habit of treating the incorporeal nature as if it were a body".[138]

While Gregory Nazianzen might be considered as having given a clear account of divine eternity, Pseudo-Dionysius (*c.* fifth century) leaves ground for misunderstanding, especially by stating that God is the cause both of eternity and of time. The common argument has been to make eternal existence one of the definitions of God, while Pseudo-Dionysius warns that it "must not be imagined that things named as eternal are simply co-eternal with God, who precedes eternity".[139] Further, the author says that both time and eternity are predicates of God, since "he is the cause of all time and eternity".[140] In Pseudo-Dionysius' view, eternity is a term frequently applied to "something very ancient or, again, to the whole course of earthly time".[141] Put in this manner, the appropriate term seems to be "everlasting". At the same time, in a typically Platonic and Neoplatonic language, Pseudo-Dyonisius defines eternity as "the measure of being",[142] "the home of being"[143] while time "has to do with the process of change"[144] and is "of the things that come to be".[145] On this, as on other issues, Pseudo-Dionysius seems to be rather unclear, but he does a lot to promote the doctrine of divine unity, which has become a common argument in the timeless view of eternity. God, according to Pseudo-Dionysius, "is one in an unchanging and transcendent way",[146] he is "eternal, absolutely perfect and always the same".[147]

With John of Damascus (*c.*675 – *c.*749), in what is probably the first *summa*, the influences of the above mentioned authors on the problem discussed here are easy to trace. In the fashion of Basil, he speaks of the seven ages of the world, while the eighth is that to come.[148] In the spirit of Gregory Nazianzen and in direct

136 Ibid.

137 Ibid., p. 251.

138 Ibid., p. 249.

139 Pseudo-Dionysius, *The Complete Works*, p. 121.

140 Ibid.

141 Ibid., p. 120.

142 Ibid.

143 Ibid., p. 121.

144 Ibid., p. 120.

145 Ibid., p. 121.

146 Ibid., p. 66.

147 Ibid., p. 116. See also p. 128.

148 St. John of Damascus, 'The Orthodox Faith' in *Fathers of the Church*, vol. 37: *St. John of Damascus: Writings* (New York, 1958), p. 204.

reference to Gregory's 'Oration 29', he tries to define the term "age" in an attempt to distinguish between time and eternity. One of the meanings of "age" is that "which is neither time nor any division of time measured by the course and the motion of the sun – that is to say, made up of days and nights [...] This kind of age is to eternal things exactly what time is to temporal things".[149] It is in this way that St. John defines one of the attributes of God and namely his eternity. At the same time, while the divine being is "uncontained", he contains and sustains all things,[150] including the whole of time: "And there is His distinctly seeing with His divine, all seeing, and immaterial eye all things at once, both present and past and future, before they come to pass".[151]

It seems appropriate to conclude the exposition on the Greek tradition on divine eternity with John of Damascus, who spells out almost word for word the Scholastic concept of *aevum*. It can be noticed that, even though some of the medieval texts give grounds for misinterpretation, on the whole, the reigning concept both in the West and the East was that of a timelessly eternal God.

Theology through Liturgy

So far I have been suggesting that the simultaneous representation of different aspects of the image on the same picture plane is informed by the concept of God's simultaneous existence, which is an aspect of the timeless view of divine eternity.

A question which might be legitimately asked is the one relating to the nature of the connection which might link the concept to the artistic practice. How was conceptual information transferred to a non-conceptual level? Is it suggested that "reverse perspective", as redefined, is a highly conscious expression of a theological doctrine? Is it actually implied that Eastern Orthodox Christians, both icon-painters and beholders, throughout the centuries have been familiar with the intricacies of the doctrine and undertook the icon as a visual witness to it?

The last two questions should, of course, be answered in the negative. As Erwin Panofsky said in *Gothic Architecture and Scholasticism*, "it not very probable that the builders of Gothic structures read Gilbert de la Porrée or Thomas Aquinas in the original".[152] Neither is it likely that the great majority of icon-painters were conversant in the Greek Fathers of the Church and the writings of St. John of Damascus. At the same time, Panofsky's study is based on the idea that Gothic

---

149 Ibid., p. 203.

150 Ibid., p. 201.

151 Ibid., p. 203. Notice again, as with St. Basil, the suggestion of priority in time, which contradicts the timeless notion, followed by St. John of Damascus in the rest of the passage.

152 Erwin Panofsky, *Gothic Architecture and Scholasticism* (Latrobe PA, reprint 2005, first in 1951), p. 23.

architects "were exposed to the Scholastic point of view in innumerable other ways".[153] In other words, it is suggested that while artists may not be explicitly aware of the conceptual meaning, works of art can nevertheless express unconscious attitudes that both artists and audience share. Whatever criticisms Panofsky's essays may have incurred, it seems that this general point, even if not its concrete elaboration, carries some conviction. The following section of this chapter will be concerned with Eastern Orthodox liturgy in its role of transmitting information from the level of "high" theology to the level, at which it addresses the average Christian believer. As Hugh Wybrew says, "the dogmatic formulations which were eventually accepted by the greater part of the Church, were soon reflected in the liturgical texts".[154] In this way, one means by which iconographers and the viewers of icons were exposed to the doctrine of timeless eternity will be considered.

Thus, the comparatively few modifications[155] of Eastern Orthodox liturgical texts which are still in use bear witness to the relatively conservative character of Eastern Orthodox theology. There is a full English translation of the corpus of Eastern Orthodox liturgical texts for the fixed feasts of Christ and Mary, published in 1969.[156] I will be using a shorter selection of texts, read on Lent, the Holy Week and Easter.[157]

If we subscribe to a timeless interpretation of divine eternity we can find an abundance of textual evidence in liturgy. Our view would be one of the possible ways to understand the bizarre but persistent use of the present tense, when referring to events that happened in the past. The use of the present tense must surely imply something more than "hope in the future".[158]

Remaining with the timeless view, let us consider the following extracts from Eastern Orthodox liturgical services: "The prophet's words are fulfilled;

---

[153] Ibid.

[154] Hugh Wybrew, *Orthodox Feasts of Mary and Christ: Liturgical Texts with Commentary* (London, 1997), p. 8.

[155] The conservative character of Eastern Orthodox liturgy, in comparison to other Christian liturgical traditions, has often been observed by scholars. According to Frank Gavin, "substantially the Liturgy of the East is that of Justinian's days" (i.e., the 6th c. AD) and it has suffered "few modifications" since (Frank Gavin, 'The Eucharist in the East and West' in W. K. Lowther Clarke (ed.), *Liturgy and Worship* (London, New York, 1932), pp. 123–4). Hugh Wybrew believes that "inherent conservatism" has been evident since the eighth century and there has been "little change" since the time of Nicholas Cabasilas (14th c.) and Symeon of Thessaloniki (15th c.), two major writers on liturgy (Hugh Wybrew, *The Orthodox Liturgy: The Development of the Eucharistic Liturgy in the Byzantine Rite* (Crestwood, New York, 1990, first in 1989), p. 173). Wybrew's view seems much better balanced, while it still points at the relatively few changes in the Eastern Orthodox liturgical tradition.

[156] See Orthodox Eastern Church, *The Festal Menaion*, tr. Mother Mary and Archimandrite K. Ware (London, 1969).

[157] From Wybrew, *Orthodox Feasts of Mary and Christ.*

[158] Cullmann, *Christ and Time*, p. 89.

for tomorrow our Lord *is* born"[159] (Christmas, Vespers service, 24th December), "Lord, you *have been* born"[160] (Christmas, Matins service, 24th December), "The Virgin *today gives* birth to him who is above being"[161] (Christmas, Matins service, 25th December), "Christ *is* baptized, and *come up* from the water, with him *raising up* the world"[162] (Epiphany, Great Compline service, 6th January), "You *have shown* yourselve *today* to the whole world, and your light, Lord, *has shone* on us"[163] (Epiphany, Matins service, 6th January). The list of examples could be much longer, but it is clear at this point that a strictly linear conception of time, commonly believed to be introduced by Judaism and sustained by Christianity,[164] could not account for the prevalent use of the present tense and expressions like "today" in liturgical texts. It is worthwhile noticing that the same applies to the Roman Catholic rite, where the Latin "*hodie"* (i.e., today) is used as much as in the East. An example is the antiphon in the responsorial psalm at the Midnight Mass of Christmas: "*Today* a saviour has been born to us; he is Christ the Lord". The Exultet sung at the Easter Vigil has a refrain: "This *is* the night", which conveys the same sense of making present the commemorated event. Traditional Anglican liturgical texts make less, if any, use of this idea, though it is to be found among modern Anglican theologians. The emphasis on remembrance, as opposed to making present, is common among Evangelicals, for instance, and in cases like that we have a straight-forward linear conception of temporality.

If Christian time is linear in the strict sense of the word then the birth of Christ has the character of the "once for all" event. How are we to understand then claims made in Eastern Orthodox, as well as Catholic, liturgical texts that the Virgin "today gives birth to him", that he is to be born "tomorrow", while "today" he shows himself, etc.? The question relates directly to the doctrine of Real Presence, which is essential to the teaching of the Eastern Orthodox Church and which divided the Church in the West during the Reformation. It is not necessary to go into the details of one of the most complicated and debated theological problems. Suffice it to say that, if in some way Christ is really present in the elements of

---

159 Wybrew, *Orthodox Feasts of Mary and Christ: Liturgical Texts with Commentary*, p. 51; all the italics in this paragraph are mine.

160 Ibid., p. 52.

161 Ibid., p. 57.

162 Ibid., p. 73.

163 Ibid., p. 74.

164 A dominant view is the one that sees the Jewish notion of time as strictly linear and opposes it to the cyclical one of the Greeks. See, for example, Gustav Hölscher, *Die Ursprunge der judischen Eschatologie* (Giessen, 1925), p. 6; Nikolay Berdyaev, *The Meaning of History* (London, 1945, first in 1936), p. 29; Stephen Gould, *Time's Arrow, Time's Cycle: Myth and Metaphor in the Discovery of Geological Time* (Cambridge, MA, 1987), p. 11.

bread and wine,[165] we need to re-examine our notion of the Christian linearity of time and its "once for all" character, which have been accepted as "the very essence of Christianity".[166]

There has been, in fact, a tendency in scholarship to overcome the strict opposition between the linear and cyclic conceptions of time. Some authors have suggested that cyclic conceptions of time, associated with certain cultures actually admit of an element of linearity.[167] The same is true of linear conceptions of time, which on closer inspection reveal some sense of circularity.[168] In the Old Testament Solomon is said to have held the view that "There is nothing new under the sun" (Eccl.1:10), while the statement "What has been, already exists, and what is still to be, has already been, and God always seeks to repeat the past" (Eccl.3.15) clearly leads to questioning the thesis of the strict linear conception of time in the Bible.

In his book *Categories of the Medieval World* Aron Gurevich maintains that a combination of the two main conceptions of time could be found in various forms in different cultures. The problem lies not in their opposition but rather in attempting to see "how these two ways of perceiving the flow of time [are] correlated".[169] As Gurevich says, "festivals and rituals form the ring connecting

---

165 The doctrine of the Real Presence, in Aristotelian terms, refers to a change of "essence", but not of the "accidents" of bread and wine. Hence, the Western term "transubstantiation", i.e., change of substance, of essence.

166 Whitrow, *Time in History: Views of Time from Prehistory to the Present Day*, p. 48.

167 On linearity within the overall cyclic temporal consciousness of the ancient Greeks see Samuel Brandon, *Time and Mankind* (London, 1951), p. 33. Arnoldo Momigliano suggests that Greek historians, on the whole, unlike philosophers, were not as much influenced by the cyclic conception (Arnoldo Momigliano, 'Time in Ancient Historiography', *History and Theory*, 5 (1966), Suppl.6: 'History and the Concept of Time', 10). According to Willem der Boer this was even more evident with Roman historians, who were actually receptive to the concept of linear time (Willem der Boer, 'Greco-Roman Historiography in Its Relation to Biblical and Modern Thinking', *History and Theory*, 7/1 (1968): 72). Ludwig Edelstein maintains that for some Greek thinkers time was non-cyclical (Ludwig Edelstein, *The Idea of Progress in Classical Antiquity* (Baltimore, 1967), p. XXI), an opinion shared by Richard Sorabji in his overview of theories of time in Greek philosophy (Sorabji, *Time, Creation and the Continuum*) and by James Barr (James Barr, *Biblical Words for Time* (Naperville, IL, 1962), p. 137).

168 André Neher goes as far as to suggest that time for the early Hebrews was cyclic (André Neher, 'The View of Time and History in Jewish Culture' in Gardet (ed.), *Culture and Time* (Paris, 1976). James Barr believes that a cyclic view could sometimes be traced in Jewish thought (Barr, *Biblical Words for Time*, p. 137). That both conceptions are present in the Hebrew tradition is also the opinion of Rudavsky (Rudavsky, *Time Matters: Time, Creation and Cosmology in Medieval Jewish Philosophy*, p. 4. On the same problem in relation to Zoroastrianism see Whitrow, *Time in History*, pp. 34–5.

169 Aron Gurevich, *Categories of the Medieval World* (London, Boston, 1985, first in Russian in 1972), p. 30.

these two concepts of time" and so the merging of cyclic and linear temporality is especially noticeable in the ritual of the Church. The Eucharistic sacrifice in particular is pregnant with a notion of time which overcomes the opposition between linear and cyclic.

Time in the Eucharist is linear in the sense that Christ's sacrifice was a unique phenomenon in history that happened once at the beginning of the Christian era. Moreover, in liturgy the main context is supplied by a prophetic look into the future, towards "the end of days" and God's world to come. The mental thrust is towards what is to come at the very end of time. At the same time, liturgical time is cyclic in the sense of *re-presenting*, making present the Incarnation and Resurrection during each new Eucharist. It is not, however, endlessly or eternally recurrent in the strict sense of that notion (since the line has a beginning and an end and so could not be eternal). The event that the Eucharist is concerned with (i.e., the Incarnation and Resurrection) is recurrent in history through the Eucharistic ritual. Thus, liturgy is not just an enactment of the Incarnation and Resurrection, but it makes present these events of sacred history in a genuine sense. If this view is problematic in scientific terms, it could make sense in a mystery that happens under the action and with the participation of an eternal being, who is outside time and its laws.

It seems at this point that remaining at the level of the linear versus cyclic dichotomy would be unproductive. We can shift the emphasis, however, and see liturgy as an advent of eternity into time. In this case, liturgical time will be defined from the standpoint of a timeless eternity, of a simultaneous divine existence, containing all time in itself. In these terms, liturgical texts, as the ones quoted earlier, gain a particular meaning as expressions of the concept of timeless eternity. Conceptual difficulties are by no means done away with, as liturgy actualizes what, according to Kierkegaard, is "the paradox" at the heart of Christianity. The Incarnation of the Word is at once "a historic event of the past" and "an event which never passes".[170] This is especially pronounced in the Eastern Orthodox rite and particularly in the "daring Eucharistic realism"[171] of the liturgy of St. John Chrysostom (d. 407). It is this "[paradoxical presence]"[172] which defines the parameters of liturgy which are meaningful only when we become aware of the nature of the being which we celebrate.

Alexander Schmemann, one of the foremost authorities on Eastern Orthodox liturgical studies, suggests that the nature of liturgy, the relationship between the "once for all" and the abiding presence, should be understood "first of all in the possibility of the conquest of time, i.e., the manifestation and realization

[170] George Florovsky, 'Introduction' in Orthodox Eastern Church, *The Festal Menaion*, p. 28.

[171] Ibid., p. 35.

[172] Ibid.

(within [the] Sacrament) of a past event in all it supra-temporal, eternal reality and effectiveness".[173]

Thus, we come back to the theme mentioned in Chapter 1 in relation to Kalokyris's article 'Byzantine Iconography and Liturgical Time'. To recapitulate, the author suggests that time in the icon is liturgical in the sense of "condensed"/ "concentré". The "condensed" nature of liturgical time corresponds closely to the simultaneity of timeless eternity. Let us consider the following passage from Evangelos Theodoron's *The Instructive Value of the Triodion* (in Greek):

> "Normally in worship, time ceases to exist in the form of past, present and future, and is changed into a mystical experience in which, while eternity is lived in the present, things of the past and of the future and even eschatological things – that is, prehistory and the main stages of the redemptive work of Christ, as well as the salutary gifts extending to the last days which followed from him – are condensed and experienced mystically as something living and present before our eyes."[174]

Liturgical time is defined in similar terms in Mark Siotes's *History and Revelation According to the Science of the New Testament* (in Greek). In Siotes's opinion, in worship "every temporal sign of the Lord's saving work is re-lived through surmounting the concept of time".[175]

Recent scholarship lays a particular stress on the timeless aspect of liturgy. In his article on 'Time' Gerald Loughlin remarks that "through repetition the passage of time is overcome and its creator glorified" and thus, "from the perspective of eternity" Christ's Second Coming is also "the first, because all of his life is simultaneously embraced in the eternity of God".[176] We already noticed the conceptual problems, in a way endemic to the timeless concept of eternity and evident in our describing it in terms, imbued with the notion of temporality. In liturgical studies such terminology, as we saw, often makes use of the conceptions of "linear" and "cyclic" time. What we have instead is the paradoxical advent of a *timeless* eternity into *time*. In this sense, the Eucharist "eternalizes time",[177] i.e.,

---

173 Alexander Schmemann, *Introduction to Liturgical Theology* (London and Portland, Maine, 1966, first in Russian in 1966), p. 35.

174 cited from Konstantinos Kalokyris, 'Byzantine Iconography and Liturgical Time', *The Eastern Churches Review*, 1 (1966–1967): 83.

175 *op. cit.*

176 Loughlin, 'Time' in Hastings, Mason and Pyper (eds.), *The Oxford Companion to Christian Thought*, p. 707. See also Joseph Ratzinger, *The Spirit of the Liturgy* (San Francisco, 2000, first in German), especially pp. 60 ff and pp. 92 ff; J. Marsh, 'Christian Worship: Human and Divine Transcendence of Time' in John Booty (ed.), *The Divine Drama in History and Liturgy* (Alison Park, PA, 1984), pp. 123–31.

177 Loughlin, 'Time' in Hastings, Mason and Pyper (eds.), *The Oxford Companion to Christian Thought*, p. 709.

raises it to the level of eternity, whereby it loses its very temporal characteristic and becomes something categorically different. Hans Belting touches upon the heart of the problem when he put the question: "in what way liturgy, and through liturgy, religious experience [...] diverted people from the perception of time and space which liturgy did not cease to counteract?".[178] I cannot pretend to answer this question, but what I find directly relevant to our purposes is the idea that liturgy did bring across a sense of transcending time.

Theology through the Arts

The previous section was concerned with the role of liturgy as a means of transmitting ideas, worked out in theology. We focused particularly on how this happens through the highly specific use of liturgical language. In this sense, we stayed within a linguistic and conceptual domain that both liturgy and theology share, so long as both put concepts into language, even though in a different way.

There is, however, another way that we acquire implicit knowledge of theological notions, namely by way of the arts (music, painting, etc.). This is a phenomenon much more difficult to account for, since we have to deal with processes, partaking of allegedly disparate domains of consciousness – the conceptual, logical and the non-conceptual, intuitive.

Panofsky's essay *Gothic Architecture and Scholasticism*, which was mentioned, belongs to a group of studies in German on the relationship between Gothic art and Scholastic theology which use this particular comparison as a case study for the more general problem of the relationship between art and theology. In the nineteenth century, the German writer Ferdinand Piper coined the term "monumental theology"[179] to describe the link between Gothic architecture and theology and to claim more broadly that artistic monuments are just as important sources for the study of theology as the texts which are usually considered.[180]

Piper's idea that theology can, in some way, be accessed via art has been re-sounded in recent years. Interesting research has been conducted in the field of the so-called "theology through the arts",[181] which is based on the belief in a possibility of mutually fruitful interaction between the conceptual and the non-conceptual, between theology and art. In particular, the arts are seen to be able to disclose or inculcate theological ideas, without, however, providing ultimate theological

[178] Hans Belting, 'An Image and Its Function in Liturgy', *Dumbarton Oaks Papers*, 34 and 35 (1980–1981): 16.

[179] In his *Einleitung in die monumentale Theologie* (1867).

[180] See Jeffrey Hamburger's discussion of Piper in Jeffrey Hamburger 'The Place of Theology in Medieval Art History', in Jeffrey Hamburger (ed.), *The Mind's Eye; Art and Theological Argument in the Middle Ages* (Princeton, 2006), pp. 18–19.

[181] For example, the ongoing project 'Theology through the Arts' under Dr. J. Begbie at Cambridge University and the University of St. Andrews.

paradigms.[182] I would like to mention some of the ideas relating to notion of "theology through the arts" in view of the subject under my attention. I will be suggesting that the definition of "reverse perspective" accepted here could actively contribute to the understanding of the theological doctrine of timeless eternity. Naturally I do not mean to maintain that a "reverse perspectival" image can in any way directly impart conceptual information on this or any other theological notion. I believe, however, that the two – "reverse perspective" and timeless eternity – share what Martin Kemp calls an analogous "structural intuition". Kemp's term refers to structures that "are both those of intuitive processes themselves and those of external features whose structures are being intuited".[183] In this way it is possible to see "reverse perspective" as creating a structure in concrete terms which acts visually in a way that is analogous to the logical structure of the doctrine of timeless eternity. Thus, the "simultaneity" of the "reverse perspectival" image becomes a structural intuition of the concept of simultaneity, growing out of timelessness. In this case, while the concept of simultaneity can be analyzed in logical terms, the structure underlying the artistic phenomenon of "simultaneous planes" can be experienced only on an intuitive level. In practical terms, however, the latter can insinuate ways of handling space and time which could prove an invaluable alternative means to the knowledge of a doctrine which, as we saw, theology and philosophy have failed to explain satisfactorily.

In a similar vein, Jeremy Begbie speaks of "theology through music". Begbie believes that "music has the capacity to play a valuable part in exposing and interpreting many of the most significant issues at stake", as well as "advancing the contemporary discussion of them".[184] The author is specifically intrigued by ways in which patterns of sound give an idea about notions of time and eternity. For instance, the extraordinary reliance of music, in comparison to the other arts, on repetition could open new avenues for grasping the phenomenon at the heart of Eucharistic repetition,[185] in that it can prove more effective than theology in many ways, among others, by freeing us "from some views of temporality […] which have disfigured the debates".[186]

In his book *Art in Action: Toward a Christian Aesthetic* Nicholas Wolterstorff, too, believes that to understand the specific interpretation of the Lord's Supper of a Church one can read its official liturgical texts, "but it is probably better to

182 Communication from Dr. Begbie.

183 Martin Kemp, *Visualizations: The Nature Book of Art and Science* (Oxford and New York, 2000), p. 1.

184 Jeremy Begbie, *Theology, Music and Time* (Cambridge, 2000), p. 128.

185 Ibid., p. 165.

186 Ibid., p. 173. In fact, Begbie is a supporter of the everlasting view, hence it is important to notice that the repetition of music is a repetition *in time*. The same basic idea applies in the case of Wolterstorff (see next passage), whose article, mentioned earlier is among the most influential interpretations of everlasting eternity.

listen to the music they use for communion".[187] For, "from a characteristically lugubrious tone, for example, one learns a lot".[188] The author even claims that "one should be able to infer the general character"[189] of people's convictions from their art, the reason for this being that "human beings in deep, unconscious ways seek fittingness between their fundamental beliefs and the character of the art they use".[190]

The connection between the visual arts and theology has a much more dramatic history. Any iconoclastic movement within the Church actually implies a fear that images can teach ideas which run contrary to the official teaching. Thus, in reference to Byzantine Iconoclasm, Charles Barber speaks of "theology through images", i.e., the notion that ideas were "carried not only by text but also by the image in the icon itself".[191] According to Barber, "the Iconoclastic dispute had a very precise point of origin, namely the 82nd canon of the Quinisext Council, 691–2",[192] which actually expressed the belief that "how one chooses to show the Christian God [as Lamb or as man] has theological implications".[193]

The belief in the power of images to impart theological ideas explains, at least partly, the violence of iconoclastic outbreaks that may look startling and even incomprehensible at first glance.[194] Thus, Robin Jensen says that "visual art often serves as a highly sophisticated, literate, and even eloquent mode of theological expression".[195] In her book *Understanding Early Christian Art*, the author studies and finds parallels between motifs and subjects in early Christian visual art and in Christian literature and liturgy at the time. In her analysis Jensen claims that there is a mutual dependence of verbal and visual modes of religious expression. The role of visual imagery is not passive, i.e., just serving and illustrating texts

---

187 Nicholas Wolterstorff, *Art in Action: Toward a Christian Aesthetic* (Michigan, 1980), p. 187.

188 Ibid.

189 Ibid., p. 188.

190 Ibid., p. 187.

191 Charles Barber, *Figure and Likeness: On the Limits of Representation in Byzantine Iconoclasm* (Princeton and Oxford, 2002), p. 53.

192 Ibid., p. 40.

193 Ibid, p. 54. The text of the canon reads: "Therefore, so that what is perfect may be depicted, even in paintings, in the eyes of all, we decree that the Lamb of God who takes away the sin of the world, Christ our God, should from now on be portrayed as a man, instead of ancient lamb, even in icons; for one might be led to the memory of his life in the flesh, his passion and his saving death, and of the redemption which thereby came to the world".

194 We should keep in mind that iconoclastic attitudes, often violent in nature, are not uncommon in modern culture, either. On that see David Freedberg, *The Power of Images: Studies in the History and Theory of Response* (Chicago and London, 1989), pp. 406–31.

195 Robin Jensen, *Understanding Early Christian Art* (London and New York, 2000), p. 3.

but rather it "paralleled, commented upon, and expanded texts".[196] The active function that Jensen claims is possessed by visual art is particularly conspicuous in cases when images actually precede texts and the latter come about to provide a commentary on already existing pictorial symbols.

While the topic of theology through the arts has been receiving increasingly more attention in recent literature it ultimately goes back to a theme prominent in Neoplatonism, especially in Plotinus and popularized in Christian theology by Pseudo-Dionysius, concerned with the non-conceptual knowledge imparted by images. The interesting question, however, is what makes it possible for an image to implicate information which is analyzable in logical terms. Michael Psellos, one of the outstanding intellectual figures in eleventh century Byzantium, seems to find the explanation in the "inspiration" of the artist, which is similar to that of the evangelist, i.e., an inspiration coming from God. In his *ekphrasis* on an icon of the Crucifixion, Psellos maintains that the painter had shown the religious paradox of Christ's dual nature, both "alive" and "human".[197] In his commentary on Psellos's text, Robin Cormack says that in Byzantium "icons were accepted as a mode through which one reaches closer to an explanation of God than any verbal definition could ever do".[198] The question, however, still remains – why does the visual become an appropriate medium for the inspired icon-painter?

This brings us back to Kemp's "structural intuitions" or "deep structures operating at a pre- or subverbal level".[199] In this context it is useful to consider the relationship between painting and consciousness. James Elkins discusses this problem in relation to ancient Indian sources[200] in which the Pali word "*citta*" refers both to painting and to consciousness. As the author notices, the implication here is that painting depicts the world in a manner similar to the one in which the mind depicts the world and that the structure of painting is "analogous to structures of consciousness".[201] In this sense, painting becomes "a demonstration of consciousness".[202] It is exactly this structural analogy between painting and consciousness that makes possible the appearance of "structural intuitions" in concrete, visual terms. In his recent lecture 'Structural Intuitions in Art and Science' (16th December 2002) Kemp goes back to the problem in *Vizualizations*. There is a fundamental analogy between "structures in our brains" and "structures 'out there'", i.e., those which underlie processes in nature. Thus, human beings

---

196 Ibid., p. 5.

197 See Doc. 28 in the Appendix to Hans Belting, *Likeness and Presence* (Chicago and London, 1994), pp. 528–9.

198 Robin Cormack, *Painting the Soul* (London, 1997), p. 112.

199 Kemp, *Visualizations: The Nature Book of Art and Science*, p. 1.

200 In particular, see Coomaraswamy, Ananda, 'An Early Passage on Eastern Painting', *Eastern Art,* 3 (1931): 218–9.

201 James Elkins, *On Pictures and the Words That Fail Them* (Cambridge, 1998), p. 202.

202 Ibid.

derive a sense of satisfaction in discovering patterns of order in the chaos of their surroundings. Actually, vision operates by intuitive mechanisms which impose structural configurations on what otherwise would be an unmanageable and chaotic array of visual impressions. In this way, for instance, artists, as well as scientists, visualize their objects by "drawing out certain aspects of geometrical order". The principle is that structured visualizations are made depends on the analogy between our mental structures and the underlying structural configurations of objects that confront us.

Medieval and Renaissance prayer books and handbooks on meditation frequently touch on this problem. Michael Baxandall brings to attention an Italian handbook, entitled *Zardino de Oration* (The Garden of Prayer), written for young girls in 1454. A characteristic passage from the text reads: "The better to impress the story of the Passion on your mind, and to memorise each section of it more easily it is helpful and necessary to fix places and people: a city, for example, which will be the city of Jerusalem – taking for this purpose a city that is well known to you".[203] What Baxandall calls the painter's "external visualizations" and the public's "interior visualizations" is well exemplified in the *Spiritual Exercises* of St. Ignatius Loyola (1491–1556). The first stage of the five proposed exercises is "composition, seeing the place" and consists in the following:

> "In contemplation or meditation on visible things, as in contemplating Christ our Lord, who is visible, composition will be to see by the eye of the imagination a physical place where that thing is found which I wish to contemplate. By a physical place I mean, for example, a temple or mountain, where Jesus Christ or Our Lady is found, according to that which I wish to contemplate".[204]

In other words, in the form of meditation of St. Loyola a mental image of a specific place at a specific time is produced and a past event appears under contemporary guise.

The examples just cited refer to one type of structural intuition which is very different from the one of Eastern Orthodox images. Among the latter's most typical features, as shown, is the resistance to the specificity of place and time. It is in this sense that they are "timeless". In visual terms, this corresponds to the non-specific locations in icons and the adherence to the Platonic Idea underlying the represented objects. Accordingly, we see, for instance, a table of a regular quadrangle shape with four legs of equal length in a non-identifiable room or undescribed space.

The sort of vision, implied by such images, is referred to in the writings of Meister Eckhart (*c.*1260–1327). Eckhart actually comes close to formulating a position, similar to the hypothesis proposed in this chapter without, of course,

---

[203] Michael Baxandall, *Painting and Experience in Fifteenth Century Italy* (Oxford and New York, 1972), p. 46.

[204] I am using Freedberg's translation from the Spanish in Freedberg, *The Power of Images*, p. 179.

using modern terminology as "reverse perspective", "simultaneous planes", etc. In a manner, he reinterprets St. Augustine's notion of "intellectual vision", which was highly influential throughout the Middle Ages.[205] According to Augustine, "intellectual vision", in contrast to the "corporeal vision" of the bodily eye and the "spiritual vision" of dreams and imagination, bestows on man the possibility of the perception of divine truth.[206] Thus, the saints in the City of God by exercizing this mode of vision (which is obviously not "vision" in the strict sense of the word), see "with an extraordinary power of sight".[207] Meister Ekhart goes one step further by maintaining that it is artistic images that are constructed according to and reveal the principle of Augustine's "intellectual vision". This idea was not at all implied by Augustine himself. At the same time, it is not surprising that it is within the tradition of mystical theology, with its inherent distrust in the conceptual knowledge of God, that we find attempts to approach the divine through non-conceptual modes, bordering on what we could call "the aesthetic". What is aimed at is, too, a form of knowledge, but the means are different.

So, Meister Eckhart starts from the premise that "anyone content with what can be expressed in words – God is a word, Heaven is a word – is aptly styled an unbeliever".[208] With the help of art, though, we can go beyond the word and even "see" things as God sees them. This is because "art amounts, in temporal things, to singling out the best",[209] i.e., the most essential. In other words, the image of an object presents us with its "form"/ "type". This is how divine vision works as well and to imitate it constitutes the ultimate aim of man in the world. As Meister Eckhart says, "to have all that has being and is lustily to be desired and brings delight; to have it all at once and whole in the undivided soul and that in God, revealed in its perfection, in its flower, where it first burgeons forth in the ground of its existence [...] that is happiness".[210] The author resorts directly to the concept of eternity when he describes this sort of vision – of God and, even though to a lesser extent, of art – as seeing things "*sub specie aeternitatis*".[211] Ananda Coomaraswamy, whose chapter on Eckhart I have been following here, summarizes the author's position as "(to see) all things in all their dimensions apart from time and space as the single objects of its [God's and, by extension, the

---

205 For a useful discussion of Augustine's position see, Cynthia Hahn, '*Visio Dei*: Vision in Medieval Visuality' in Robert Nelson, (ed.), *Visuality before and beyond the Renaissance* (Cambridge, 2000), pp. 169–97.

206 St. Augustine, *De genesi ad litteram* (The Literal Meaning of the Genesis) (New York, 1982), vol. 2, Book XII.

207 Cited in Hahn, '*Visio Dei*: Vision in Medieval Visuality', p. 170.

208 Meister Eckhart, *Meister Eckhart* tr. C. Evans (London, 1924–1931), p. 339.

209 Ibid., p. 461.

210 Ibid., p. 82.

211 Ibid., p. 47.

artist's] vision, not turning from one thing to another but seeing without light, in a timeless image-breaking light".[212]

Thus, it could be noticed that Meister Eckhart understands art very much within our framework of icon art. On this, he is close in spirit to mystical theologians before him such as Pseudo-Dionysius, as well as to later Eastern Orthodox, especially Russian, thinkers, strongly influenced by mysticism.[213]

A major implication of the "theology through the arts" is that the division between the conceptual and non-conceptual mode is frequently taken too far. Both have a relationship to consciousness that makes it possible that they share analogous deeper structures. Thus, it is worthwhile considering that the structure of a visual image might work in analogous ways to certain aspects of a proposition, as made in theology. The "simultaneity" of the "reverse perspectival" image may be seen as containing in the very principle of its spatial organization a visual code to the nature of a simultaneously existing, timeless deity.

The Present Hypothesis in Context

I cannot cite an authority that directly supports my hypothesis. At the same time, I believe that my main idea and approach to the problem may be seen in the light of a certain trend of thought in philosophy and theoretically oriented art history, which illustrates the transformation from a religious to an aesthetic attitude. Thus, the theological concept of simultaneity is appropriated to describe aesthetic experience.

Leibniz and the Way "God Sees Things"

The closest I have come to a formulation similar to mine comes from a rather unexpected quarter, in the sense that it is by an author completely outside the Eastern Orthodox tradition of thought and artistic practice. In a letter to the Jesuit theologian and mathematician Des Bosses, dated February the 5th 1712, Leibniz claims that there is a fundamental distinction between "the appearance bodies have with respect to us and with respect to God".[214]

The differences and analogies between the divine and human points of view are part of Leibniz's overall concern with the relationship between man's nature and God's nature, which formed a major trend in the thought of the time.

---

212 Ananda Coomaraswamy, 'Meister Eckhart's View on Art' in his *The Transformation of Nature in Art* (Cambridge, MA, 1934), p. 93.

213 On this see Vladimir Lossky, *The Mystical Theology of the Eastern Church* (Cambridge and London, 1957). It is important to notice that, according to Lossky's thesis, all Eastern Orthodox thought is essentially mystical.

214 Gottfried Leibniz, *Philosophical Essays* (eds.) Roger Ariew and Daniel Garber (Indianapolis, Cambridge, 1989), p. 199.

Analogies between divine and human nature had been a prominent leitmotif among Renaissance thinkers. It was during the Enlightenment, however, that they formed part of a consistent philosophical concern. In the seventeenth and early eighteenth centuries theses about divine and human properties became, according to Edward Craig, "less an idea than an attitude".[215]

God's vision is characterized in the following way in the letter to Des Bosses:

> "God sees things exactly as they are in accordance with geometrical truth, although he also knows how everything appears to everything else, and so he eminently *contains in himself all other appearances*" (the italics are mine).[216]

The notion that God has a knowledge of all possible appearances of objects, i.e., from all possible points of view, is clearly, I believe, another way of saying that divine point perception is timelessly eternal, non-spatial and non-locational in the way described above. It is important to notice that with Leibniz "possible" refers to all that is actual, i.e., already created by God but also the potential, i.e., that which could be created by God if he chooses. Hence the talk of "possible worlds" deriving from Leibniz. As Leibniz says in the *Monadology* 43, every possible substance, not only the ones singled out for creation, is represented in the mind of God by its "complete individual notion", in which every detail of the substance at every stage of its potential career is fixed. As he subscribes to the theological doctrine of God as the perfect being, God's containing and knowing all possible substances, making up the possible worlds, is an expression of his omniscience. Since God is omnipotent, too, he can bring into existence any of these worlds, and, being omni-benevolent, he, of course, chose to create the best possible world. This is, in short and simplified terms, how Leibniz's argument runs. What concerns us here is that divine vision incorporates at the same time, in Leibniz's terms, all actual and possible aspects of all actual and possible objects.

After saying that "God sees things [...] in accordance to geometrical truth", Leibniz continues with the analogy and claims that to divine perception things appear like "a ground plan or geometrical representation" in contrast to the appearance revealed to human vision, which is like "a drawing in perspective".[217] While Leibniz does not refer to any of the properties we associate with "reverse perspective", one idea is of immediate interest. It is significant that Leibniz claims that objects would appear fundamentally differently according to the fundamentally different viewing positions under which they are seen – the human or the divine one. As Leibniz remarks in another letter, this time to Arnauld, from 9 October

215 Edward Craig, *The Mind of God and the Mind of Man*, p. 13. See also Nicholas Rescher, *The Philosophy of Leibniz* (Englewood Cliffs, New Jersey, 1967), especially, pp. 11–22; Anthony Savile, *Routledge Philosophy Guidebook to Leibniz and the Monadology* (New York and London, 2000), pp. 43–63.

216 Leibniz, *Philosophical Essays*, p. 199.

217 Ibid.

1687, "created minds differ from God only in degree, from finite to infinite".[218] What is also noteworthy is that he makes an explicit reference in the letter to Des Bosses to the way this is illustrated through pictorial representation.

It would be wrong, however, to put too much emphasis on the distinction between divine and human vision and nature, as Leibniz was mainly concerned in drawing analogies between man and God. The passage just quoted actually follows the statement that the universe is "made up of as many little Gods beneath this great God".[219] Thus, Leibniz is giving expression to a theme, typical of his time, which could be put in Hamlet's words, marvelling at "What a piece of work man is! […] how *infinite in faculty*! […] in apprehension, how *like a god*!" (Shakespeare, *Hamlet*, Act III, Scene 2; the italics are mine). The intriguing thing is that Leibniz goes as far as to propose a similarity between God's infinite, and so simultaneous, perception and man's. The following passage from the *Discourse on Metaphysics* 9, deserves special attention in this context:

> "One can even say that every substance bears in some sort the character of God's infinite wisdom and omnipotence, and imitates him as far as it is capable. For it expresses, albeit confusedly, all that happens in the universe, past, present or future, and this has some resemblance to an infinite perception or knowledge."

Is Leibniz suggesting that we can indeed see the world with the "eyes" of a timeless and infinite God? The passage has caused strong criticisms[220] and it is probably best to understand Leibniz as having meant that while our perceptions are confused and we are not even conscious of most of them[221], they do reflect, though on a smaller scale, the infinity of God.[222] However contentious this idea is in Leibniz, we come across it again, in a different form of course, especially with thinkers who do away with God's perspective altogether and transfer divine dimensions to certain states of human awareness and perception.

Schopenhauer and Art as a "Repetition of Eternity"

It is hardly to be expected that Schopenhauer, "who never had any place for God in his philosophy"[223] would evoke a divine point of view in any way. He did something, though, which makes him one of the most important thinkers in terms of providing a philosophical context for the hypothesis proposed in this dissertation.

---

218 Gottfried Leibniz, *The Leibniz-Arnauld Correspondence* ed. Haydn Mason (New York, 1967), p. 160.

219 Ibid., p. 159.

220 See Craig, *The Mind of God and the Mind of Man*, p. 60.

221 Ibid., p. 59.

222 Ibid., p. 60.

223 Christopher Janaway, *Self and World in Schopenhauer's Philosophy* (Oxford, 1989), p. 206.

Schopenhauer makes art of key significance in his overall metaphysical system and as such it serves a specific role, namely as a means to metaphysical truth. Thus, aesthetics, in a sense, takes the place occupied in Christian thinking by theology. Schopenhauer's notion of the non-specificity of aesthetic perception, which we will discuss below, is very similar in structural terms to the idea of God's ubiquitous perception in "reverse perspective".

It is typical of Schopenhauer's philosophy that it is art and not science that provides the path to true knowledge.[224] Most importantly, art achieves its ultimate purpose through a perception, devoid of time, space and sufficient reason.[225] Once, in the act of aesthetic contemplation, we have overcome the constraints, imposed by the *principium individuationis* (i.e., time and space) a fundamental transformation occurs both in us as viewing subjects and in the objects of our contemplation. This is how Schopenhauer interprets Spinoza's statement that "the mind is [in contact with] eternity insofar as it conceives things from the standpoint of eternity",[226] a theme familiar also from Leibniz. It is telling that "the standpoint of eternity" is, according to Schopenhauer, the one we adopt in aesthetic contemplation. It is art that "repeats and reproduces eternity" and is "concerned with that which is outside and independent of all relations, that which alone is really essential to the world, the true content of its phenomena, that which is subject to no change, and therefore is known with equal truth for all time".[227] The object of art, described in these terms, is, for a Christian, God, while for Schopenhauer it is the Will. Even at this initial stage, Schopenhauer's terminology could produce some confusion and this has been frequently noticed by scholars.[228]

Schopenhauer talks about the Will as something intrinsically evil. Will is, in Christopher Janaway's words, "a permanent principle of insatiable striving with countless phenomenal manifestations".[229] So long as Will is at the root of all misery in the world – and the world, according to Schopenhauer is miserable in general and life is not worth living – art (as well as ascetic denial) provides the only means, at least temporarily, for an escape from willing and unhappiness. At the same time, however, while "[celebrating] a holiday from the penal labour of willing" we gain "knowledge of the true essence of the world, i.e., the Idea".[230] The problem arises from the conclusion that, in Schopenhauer's terms, we get to know the world exactly as Will. It makes sense to consider at this point Michael Tanner's remark that since Will is evil and it informs the ultimate nature of things then it follows

[224] See, for example, Arthur Schopenhauer, *The World as Will and Representation* (New York, 1958, rpt.1969, first in German in 1819), vol.1, p. 177.

[225] Ibid., vol.2, p. 364.

[226] Ibid., vol.1, p. 179.

[227] Ibid., vol.1, pp. 238–9.

[228] Janaway, *Self and World in Schopenhauer's Philosophy*, (Oxford, 1989), p. 202.

[229] Ibid., p. 271.

[230] Arthur Schopenhauer, *Arthur Schopenhauer: Manuscript Remains in Four Volumes*, vol. 1, (Oxford, 1989–1990), p. 129.

that Ideas, too, "must be ideally bad".[231] It thus becomes "puzzling how a close relationship with them [the Ideas] might prove valuable or pleasurable".[232]

The fact that Schopenhauer takes Kant's concept of noumenal will and extends it to cover the world in general and all the phenomena in it has often provoked severe criticism[233] and not a few have considered it "a characteristic piece of metaphysical extravagance".[234] One way of getting out of the problem has been suggested by John Atwell and it consists in making a distinction between individual will and Will as underlying the world.[235] If Atwell's suggestion is accepted then, among other things, the subject's renunciation of will in aesthetic contemplation would be understood as a giving up "only [of] individual/egoistic will, not [of] the will *in toto*".[236] This would do away with the tension that appears in Schopenhauer between "the complete silence of the will"[237] and the activity which is presupposed in the act of casting away of the will. Tempting as Atwell's suggestion is we should be cautious about it, as it looks highly unlikely Schopenhauer would have accepted it. With all the risks of it, it might prove safer to stay within an interpretation of Schopenhauer's philosophy as a "Platonism turned sour".[238]

Whatever problems and even contradictions arise out of Schopenhauer's use of the term "will" it is clear that he is committed to his thesis that art is of value and its value consists in the ability to liberate us from the will and give us a glimpse of happiness and knowledge. It is also clear that this freedom from the will is a freedom from the *principium individuationis*. What does it mean for us to have an experience which is outside the spatial, temporal and causal framework? In other words, how do we perceive objects if our perception is not subject to the laws of space, time and causality?

To answer these questions, we must notice the implications of a major departure of Schopenhauer from Kant's position in the 'Transcendental Aesthetic'. When Kant says that time and space are necessary conditions of perception it follows that it is impossible to perceive an atemporal object. Schopenhauer, on the other

---

231 Michael Tanner, *Schopenhauer: Metaphysics and Art* (London, 1998), p. 33.

232 Ibid.

233 For example, Lukács's accusation that Schopenhauer "anthropologizes the whole of nature" (Georg Lukács, *The Destruction of Reason* (London, *c*.1980), p. 225). Janaway, though, disagrees with Lukács, saying that the majority of the manifestations of will are not instances of conscious willing (Janaway, *Self and World in Schopenhauer's Philosophy*, p. 202).

234 Tanner, *Schopenhauer: Metaphysics and Art*, p. 13.

235 John Atwell, 'Art as Liberation: A Central Theme of Schopenhauer's Philosophy' in Dale Jacquette (ed.), *Schopenhauer, Philosophy, and the Arts* (Cambridge, 1996), p. 89.

236 *op. cit*., p. 92.

237 Schopenhauer, *The World as Will and Representation*, vol.2, p. 373.

238 Janaway, *Self and World in Schopenhauer's Philosophy*, p. 274.

hand, defines aesthetic experience exactly as the perception of such atemporal objects.[239]

It becomes obvious that in Schopenhauer's view, in aesthetic contemplation a great transformation is affected both in the subject and in the object of perception, whereby they lose their subjective, individual aspects, characteristic of them in the spatio-temporal and causally-related world. The spectator is "elevated [...] above himself, his person, his willing".[240] He/she becomes a "pure, will-less, timeless subject of knowledge".[241] Its objects are no longer individual bodies but eternal Ideas. As Schopenhauer says, "that what is thus known is no longer the individual thing as such but the Idea".[242] Thus, in contemplating a horse we are brought into contact not merely with a particular horse, but with the Idea of a horse in general. In Schopenhauer's own terminology, aesthetic perception is of "pure objects"[243] by the "pure subject of knowing".[244] For our purposes, it can be maintained that both the subject and the object in the Schopenhauerian sense are part of the experience of a timeless eternity.

Schopenhauer's definition of the pure subject and object is, of course, explicitly referred to Plato.[245] Schopenhauer's pure subject stands in a direct line to the Platonic Soul, which exists eternally and, in separation from the body, knows eternal Forms. At the same time, Schopenhauer's divergences from Plato are highly illuminating. Plato's Forms can be known only by conceptual thought, while Schopenhauer's pure objects are intuited in aesthetic contemplation. This partly explains the interest that Schopenhauer held for part of later philosophy and aesthetics, dealing with art's unique way of imparting knowledge.[246] It also draws a bridge back to Plotinus and Schopenhauer at one point almost literally

239 On the difference between Kant and Schopenhauer see, for example, Janaway's detailed account in Janaway, *Self and World in Schopenhauer's Philosophy*; also T. J. Diffey, 'Schopenhauer's Account of Aesthetic Experience', *British Journal of Aesthetics*, 30/2 (1990): 140.

240 Schopenhauer, *The World as Will and Representation*, vol.1, p. 201.

241 Ibid., vol.1, p. 179.

242 Ibid.

243 Ibid., vol.1, p. 130.

244 Ibid., vol.2, p. 364.

245 On Plato's influence on Schopenhauer see, for example, Christopher Janaway, 'Knowledge and Tranquility: Schopenahauer on the Value of Art' in Jacquette (ed.), *Schopenhauer, Philosophy, and the Arts*, pp. 39–62.

246 On Schopenhauer's influence on the late Heidegger, see Julian Young, 'Schopenhauer, Heidegger, Art, and Will' in Jacquette (ed.), *Schopenhauer, Philosophy, and the Arts*, pp. 162–81. According to Young, Wittgenstein in the *Tractatus* and Nietzsche in *The Birth of Tragedy* had actually "plundered" Schopenhauerian philosophy (p. 162). On the influence of Schopenhauer on Anglo-American aesthetics since the Second World War see T. J. Diffey, 'Schopenhauer's Account of Aesthetic Experience', *British Journal of Aesthetics*, 30/2 (1990): 132–42.

repeats Plotinus when he says that "all original and primary thought takes place figuratively".[247]

The essential role that art plays in imparting knowledge in Schopenahuer is intimately connected with another "correction" of Plato. In Plato's world-view, while philosophy and conceptual thought reach to the universal, art can only know the particular. For Schopenhauer, on the contrary, what is significant in art is not "the particular but the universal in it".[248] As Julian Young remarks, "in this he [Schopenhauer] is surely right and Plato wrong. For only this view can account for the deep significance that art has, and is accepted as having, in human life".[249]

In conclusion, it appears that the metaphysical function of art is of enormous importance as it affects an in-depth transformation in man and the objects of his perception. Both the subject of aesthetic contemplation and its objects, as we saw, are raised to a level of universality and objectivity and thus pertain of eternity. Schopenhauer never tires of emphasizing this aspect. "We are no longer individual", he says, "we are now only there as the one world-eye, which looks out from all knowing beings".[250] The object we contemplate as "pure" and "timeless" subjects is a thing "plucked out from the stream of the world's course".[251]

Worringer's "Eternalization" of the Object

There are various ways in which this Schopenhauerian attitude to art has influenced art historians. A work close in spirit, and often in language, to the Schopenhauer's aesthetics is Wilhelm Worringer's *Abstraction and Empathy* (1908). In it the author considers "the possibility of taking the individual thing of the external world out of its arbitrariness and seeming fortuitousness, of eternalising it by approximation to abstract forms and, in this manner, of finding a point of tranquillity and refuge from appearances".[252] What Worringer calls "the urge for abstraction", typical of certain civilizations, finds expression in the creation of art forms, which represent the object as "[approximated] to its *absolute* value".[253]

---

247 Schopenhauer, *The World as Will and Representation*, vol.2, p. 378. The earlier English translation renders that as "all primary thought takes place in pictures" (Schopenhauer, A., *The World as Will and Idea* (London, 1883, reprint 1964), p. 141). See Plotinus, *The Enneads* V.11, where the author says that Egyptian sages thought in images, which is the original mode of human thinking and only later man began to think in concepts.

248 Ibid., vol. 1, p. 231.

249 Julian Young, 'The Standpoint of Eternity: Schopenhauer on Art', *Kant-Studien*, 78 (1987): 437.

250 Schopenhauer, *The World as Will and Representation*, vol.1, p. 254.

251 Ibid., vol. 1, p. 187.

252 Wilhelm Worringer, *Abstraction and Empathy* (London, 1953, first in German in 1908), p. 16.

253 Ibid.

To give an idea of what Worringer means under "the urge to abstraction" would involve providing a brief outline of the thesis of *Abstraction and Empathy*. This seems worthwhile as in this work there are several themes running which are of special significance for the present chapter. Some of the concerns of the author sound similar to Florensky's and really are typical of critical thought of the beginning of the twentieth century. Though not going all the way, Worringer points in a direction which bears a close affinity to the one that gave rise to the hypothesis of this chapter. We will not be concerned with a critical analysis of Worringer's theory of art, but will only follow some aspects of it which are of immediate interest here.[254]

A theme which provides the background to Worringer's book is the appeal to a relativist approach to art forms. Worringer frequently mentions his debt to Alois Riegl's theory of art in that respect. According to Riegl, the work of art is an objectification of a *Kunstwollen* or an artistic volition. Without going into detailed definitions of the concept of *Kunstwollen*[255] what is of immediate interest is that Riegl proposed it as an alternative to the Semperian materialistic theory of art.[256] The followers of Semper, by crudely interpreting Semper himself, suggested that, as Worringer puts it, the history of art is a history of ability.[257] Art progresses from primitive to modern depending on the stage of development of three factors – utilitarian purpose, raw material and techniques. Thus, the statements one often comes across that the "distortions" in icon art are due to the lack of skill of iconographers, who had not yet mastered linear perspective and other means of naturalistic representation, belong to that Semperian tradition. What Riegl was saying was that the history of art was not a matter of degrees of ability but was rather a history of *volition*. As the *Kunstwollen* changes and transforms so do the art forms which are its expression.

Worringer sides with Riegl in that relativist approach to art and maintains that "what appears from our standpoint the greatest distortion must have been, at the

---

254 For a critical study of Worringer see, for example, Paul Crowther, *The Transhistorical Image: Philosophizing and Its History* (Cambridge, 2002), pp. 12–16.

255 On Riegl's concept of the *Kunstwollen*, see Erwin Panofsky, 'The Concept of Artistic Volition' (1920), *Critical Inquiry*, 8 (1981): 7–34; on Panofsky's essay see Margaret Iversen, *Alois Riegl: Art History and Theory* (Cambridge, MA, and London, 1993), pp. 152–7; further on Riegl's concept see Otto Pächt, 'Art Historians and Art Critics – Alois Riegl', *Burlington Magazine* (1963): 190–1; Margaret Olin, M., *Forms of Representation in Alois Riegl's Theory of Art* (University Park, PA, 1992), pp. 148–55; Margaret Iversen, *Alois Riegl: Art History and Theory* (Cambridge, MA, and London, 1993), pp. 3–21; Paul Crowther "More Than Ornament: Riegl and the Problem of Style" in *Transhistorical Image*, Margaret Olin, M., *Forms of Representation in Alois Riegl's Theory of Art* (University Park, PA, 1992) On the concept of *Kunstwollen* in general and specifically in Worringer see Crowther, *The Transhistorical Image*, pp. 12–13.

256 On Semper and his followers, see Michael Podro, *The Critical Historians of Art* (New Haven and London, 1982), pp. 44–61.

257 Worringer, *Abstraction and Empathy*, London, p. 9.

time, for its creator the highest beauty and expression of his artistic volition".[258] He powerfully stresses the point that it is misleading to judge all art forms with criteria developed for dealing with classical Antique and Renaissance art. Such judgements are, according to Worringer, "absurdities and platitudes".[259] Each art form requires the working out of a system of concepts relevant to its specific nature. The theory of empathy,[260] popular at the beginning of the twentieth century, Worringer claims, can be applied with justification only to the art of classical Antiquity and the Renaissance. It falls, however, outside the sphere of non-naturalistic artistic traditions. To deal with those, the author proposes the concept of abstraction.

Worringer was specifically interested in Byzantine and also Egyptian art, as the examples he draws often show. The limited applicability of the empathy theory becomes clear in its helplessness before "a lifeless form of a pyramid" or the "suppression of life in Byzantine mosaics".[261] One thread of thought throughout Worringer's work concerns an aspect of the noticed "lack of life" in non-naturalistic art forms. The "eternalization" of the images directly implicates the suppression of time.

At this point, it becomes necessary to see how the process of eternalizing the image is realized. While Worringer's reply might not be fully satisfying as it is too general, it is important that it focuses on spatial construction. It bears witness, however, to Worringer's not being able to completely break free from certain Western notions that he set out to overcome. In citing Riegl, Worringer mentions three features that stand at the heart of non-naturalistic art. The first two directly pertain to matters of pictorial space – "the approximation of the representation to a *plane*" and "the strict *suppression of space*" (the italics are mine).[262] Both are ways of dealing with "the immense dread of space" typical of mainly ancient civilizations. However, just as with Riegl, Worringer's ideas on this issue are grounded in the typically Western notions of a "plane" versus "space" – ancient artistic traditions consciously "suppress" the latter and orient objects to the former. This argument presupposes that space is a given condition of representation and has to be negated. If we assume that space (in the Western sense of the term) is not a given, the logic would be very different.[263] Florensky is close to Worringer on this point as well, since he too, as we saw in Chapter 2, kept moving within a framework of Western art categories, the validity of which he claimed to deny.

258 Ibid., p. 13.

259 Ibid.

260 On the theory of empathy, see Harry Mallgrave and Eleftherios Ikonomou (eds.), *Empathy, Form and Space: Problems in German Aesthetics 1873–1893* (Santa Monica, CA., 1994).

261 Worringer, *Abstraction and Empathy*, p. 14.

262 Ibid., p. 21.

263 I thank Prof. M. Kemp for bringing this problem to my attention.

The third feature of non-naturalistic art, according to Worringer, concerns "the exclusive rendering of the single form"[264] and this is of central importance in view of the hypothesis of the present paper. The "exclusive rendering" comes down to the attempt at the representation of the form as a "thing in itself", which would mean the form as it exists transcendentally out of space and out of time. In other words, to render exclusively the single form would mean to eternalize it, to realize the "instinct for the thing in itself".[265] It is interesting to notice that Florensky uses strikingly similar terminology when he talks about the essence of the icon.

While empathy is associated with the concept of naturalism, its counter-pole of abstraction is paired with style. And the essence of style is very much what Schopenhauer defines as the essence of art as such. "In everyday speech", Worringer tells us, "the style of a work of art implies that which lifts the natural model into a higher sphere, in other words that trimming which the natural model has to put up with in order to be transposed into the language of art".[266] Style becomes a label for non-naturalistic art in so far as "the single thing of the external world" is consciously "eternalised" by being "[wrestled] from its temporality and unclarity".[267] The urge to abstraction consists exactly in this "snatching at the 'thing in itself'".[268]

At no point does Worringer explicitly refer to the phenomenon of "simultaneous planes" in "reverse perspective". He speaks, however, of the conscious eschewing of depth values, which were felt by ancient peoples as the "subjective clouding of an objective fact".[269] It is this desire to represent the object in its absolute value that, Florensky suggested, has lead to image-making in "reverse perspective". Florensky's notion of the "simultaneous planes", I believe, brings Worringer's argument to a logical conclusion. It comes as no surprise that Worringer's book is in Florensky's bibliography for his lectures at the Russian Academy of Artistic Sciences,[270] as there definitely seems to be a close link between the two authors.

Finally, it deserves mention that, in the context of his discussion of Byzantine art, Worringer suggests that it is possible "to deduce psychic presuppositions

---

264 Worringer, *Abstraction and Empathy*, p. 21.

265 Ibid., p. 18.

266 Ibid., p. 33.

267 Ibid., p. 37.

268 Ibid., p. 38.

269 Ibid., p. 22.

270 See Nicoletta Misler, 'Toward an Exact Aesthetics: Pavel Florensky and the Russian Academy of Artistic Sciences' in John Bowlt and Olga Matich (eds.), *Laboratory of Dreams: The Russian Avant-garde and Cultural Experiment* (Stanford, 1996), p. 131. It is interesting that Worringer's influence has been noticed on other members of the Academy, as for instance, on Kandinsky's writings on art (see Alain Besançon, *The Forbidden Image: An Intellectual History of Iconoclasm* (Chicago and London, 2000, first in French in 1994), Note 18 to Chapter 8, p. 405). Dora Vallier, though, has disagreed with this view (Dora Vallier, *L'art abstrait*, Paris, 1980).

of such an artistic volition from religion and the world view of the people in question".[271] This is exactly what I have tried to do in this chapter – to build a bridge between a theological doctrine at the heart of a religious worldview and the phenomenon of representations, intended to serve devotional functions.

Conclusion

This last chapter proposes an alternative explanation of the phenomenon of "reverse perspective". As was revealed, there are serious problems and faults with the prevailing view (Chapter 2). By proposing a new definition of "reverse perspective", the following hypothesis became possible: the simultaneous representation of different planes of the iconic image, i.e., what we called "simultaneous planes", is linked to the theological doctrine of a God, who is timelessly eternal, i.e., exists simultaneously and is not subject to spatial location. If this hypothesis is accepted as true, one of the conclusions would be that the temporal component acts as a major organizing principle in the pictorial space of icon art and more generally that time does play a significant part in the visual arts, unlike what is frequently believed (Chapter 1). Still further, the connection between the theological doctrine of timeless eternity and the pictorial space of the icon confirms the notion of the sacred image as a container of presence (Chapter 3).

The nature of the link between the artistic phenomenon of "simultaneous planes" and the theological doctrine of a timelessly eternal and, by implication, simultaneous, divine existence is revealed in a twofold manner. On the one hand, the said doctrine informs the artistic practice largely via liturgy. On the other, "reverse perspective" as defined here can provide a visual key to the logical structure, underlying the theological concept.

Finally, three authors, who have been concerned with the theme of the "eternalization" of the image in art, were discussed. From Leibniz, via Schopenhauer and to Worringer, a development has been traced in which art encroaches progressively on territory hitherto reserved for theology. Ultimately, one is led to consider the relationship between art and theology as much more complex and far richer than might appear at first glance.

---

[271] Worringer, *Abstraction and Empathy: A Contribution to the Psychology of Style*, p. 101.

Conclusion

This book was organized around several interlocking themes and a central thesis. First, it has been proposed that "reverse perspective" is an essential characteristic of icon art. Attention was paid to the fact that the principle of "reverse perspective" has been employed with astonishing persistence over the centuries and has become an almost permanent element of the make-up of the icon. The typical and "eccentric", for a Western viewer, appearance of Byzantine and Byzantining images is largely due to their treatment in "reverse perspective". In other words, according to the view adopted in the present work, "reverse perspective" is important as one of the stylistic features which allows us to set apart within the overall history of art a group of images under the label of "icon art". The criterion here is not a geographic one as, even though icons have been produced overwhelmingly in Byzantium and what Dimitrii Obolensky has called "the Byzantine Commonwealth",[1] we come across them outside these territories as well. While being the cult object *par excellence* for the Orthodox believer icons, as we saw, have been produced by and for the non-Orthodox. It is on art-historical grounds that we speak of icon art in connection to images that are predominantly, but not exclusively, associated with the Eastern Orthodox Church.

Second, while "reverse perspective" can by no means be accepted as the result of the lack of artistic mastery – a view prevalent till the end of the nineteenth century – it can be linked to a particular theological concept that was common to Eastern and Western Christianity in the medieval period. My hypothesis has been that there is an underlying structural similarity, a "structural intuition", between a certain understanding of "reverse perspective" and the doctrine of a timelessly eternal god. This idea, however, has depended on a critique of the existing understanding of "reverse perspective" and on the working out of an alternative position. Thus, I understand "reverse perspective" in terms of "simultaneous planes". A structural analogy can be drawn between this principle of "reverse perspective" and the simultaneity of a timelessly eternal god. It is in this sense that the icon represents, in a visual manner, "the way God sees the world". In simple terms, a simultaneously existing god, who is outside space (i.e., has no "point of view") and time (i.e., has no "successive vision") would perceive the various aspects of objects *all at once*.

Third, "reverse perspective" understood in the above fashion, has been accepted as an element of the form, authorized to carry the presence of the

[1] Dmitri Obolensky, *The Byzantine Commonwealth: Eastern Europe, 500-1453* (London, 1971).

prototype in the image. The construction of the sacred image becomes one of the major factors that signal presence which affects the very function and nature of the image. The icon, therefore, is different from the Kantian object of disinterested aesthetic contemplation. Aspects of its theological and philosophical significance come across in the Byzantine theology of the image, as well as in early twentieth-century Russian philosophy.

The above three points have grown from the concrete attempt to suggest an alternative understanding of "reverse perspective" which challenges the theory informing practically all serious studies on the subject so far. In the process, however, much wider general assumptions have been questioned. The new view on "reverse perspective", proposed here, depends on the notion that time plays a significant part in the organization of pictorial unity. In particular, the temporal conception has been seen as inextricably linked to the formal means employed in the construction of the icon.

Further, this principle of the organization of pictorial space turns the icon into a highly specific image – an image invested with the power of containing real presence. This could question the very distinction between subject and object which lies at the heart of aesthetics. In whatever sense we understand presence, an animated image cannot be regarded as a passive object. In a manner, when "the object stares back"[2] the relationship between viewing subject and viewed object is radically transformed.

My main objective was not to provide clear-cut solutions to the problems that were raised, but to take an initial step towards the working out of categories that describe icon art on its own terms. The three main categories that have emerged are time, space, and presence, as reflected in my title. The need for new terms and notions, and the elaboration of already existing ones, arises from the belief in the inadequacy of the existing categories of Western aesthetics to describe the fundamentally different artistic phenomenon of the icon. Time, space, and presence are thus put forward as candidates in the field of the still non-existent "Eastern Orthodox aesthetics" (where "aesthetics", too, is a problematic term as it, too, is a creation of Western thought). It has been claimed that they refer to features which are of central importance for the understanding of the icon.

At the end of the work, there is, as frequently happens, a somewhat disturbing realization of having reached only the beginning of a road. No conclusive theory has been produced on "reverse perspective" in the place of the view which we have opposed. There are many other aspects of the role of time in icon art that should be explored before we come to a satisfactory idea of its significance in the organization of the pictorial unity of the icon. Some of the questions, relating to the theology of the image seem to lead in the direction of an Eastern Orthodox aesthetics and this is yet another large and unexplored field. All these and other relating problems are of considerable importance for the appreciation

2 James Elkins, *The Object Stares Back: On the Nature of Seeing* (New York, London, 1994).

of Eastern Orthodox art and culture. I have attempted to "clear the ground" as much as possible from long-standing and deep misconceptions. My main thesis is one of the possible interpretations of the material, relying on new or reworked categories.

Sample Analysis

An Analysis of Rublev's *Trinity* Icon in the Terms, Proposed in this Book

Andrey Rublev's (1350–1425) icon of the *Holy Trinity* (Fig. 5.1) is very probably not only the best known, but also the most revered Russian image. The images painted by Rublev's hand were officially declared prototypes to be followed by iconographers by the Moscow Council of the Hundred Chapters (1551). They were believed to have miraculous powers, while Rublev himself, ever since his death has been widely considered a saint,[1] even though he was officially canonized by the Russian Church only in 1988. Even during the period of early Soviet iconoclasm, Rublev's works retained their exclusive status and were included in Lenin's list of monuments worthy of preservation (in the 1918 edict 'On Monumental Propaganda').

Volumes have been written about the *Trinity* icon, especially since the first cleaning and the removal of the *oklad* (cover) in 1905–1906.[2] I will be mainly concerned with writings by Russian authors, which constitute the large bulk of criticism. Rublev's work provided an unprecedented focus for the revived interest in Russian medieval art and culture, which had been an important dimension of the intellectual history of Russia since the second half of the nineteenth century. Rublev's image, therefore, becomes an appropriate candidate for a sample analysis at the end of this book, which has been largely concerned with working out new categories with which to the art of the icon. How adequate are these categories and how meaningfully can they describe a particular image? Can they add anything illuminating to the understanding of an image like Rublev's which has been interpreted and even overinterpeted?

It will be suggested in the following analysis that Rublev's work is an *icon* in the sense we have accepted. Further, it is seen as a successful manifestation of *theology through art* which gave it officially the status of a *prototype*. The theological dogma of the Trinity is insinuated by means of the elaboration on the principle of *simultaneity*, underlying *reverse perspective*. As a result, the image

---

[1] Florensky, for instance, frequently refers to Rublev in this way.

[2] There are several useful anthologies – Part 3 in Victor Lazarev, *Andrey Rublev i ego shkola* (Andrey Rublev and His School), (Moscow, 1966), pp. 75–104, Gerol'd Vzdronov, (ed.), *Troitsa Andreia Rubleva: Anthologiia* (The Trinity of Andrey Rublev: An Anthology), (Moscow, 1981). Also, see the outline of texts in Russian in Lindsey Hughes, 'Inventing Andrey: Soviet and Post-Soviet Views of Andrey Rublev and His Trinity Icon', *Slavonica*, 3/2 (2003): pp. 83–91.

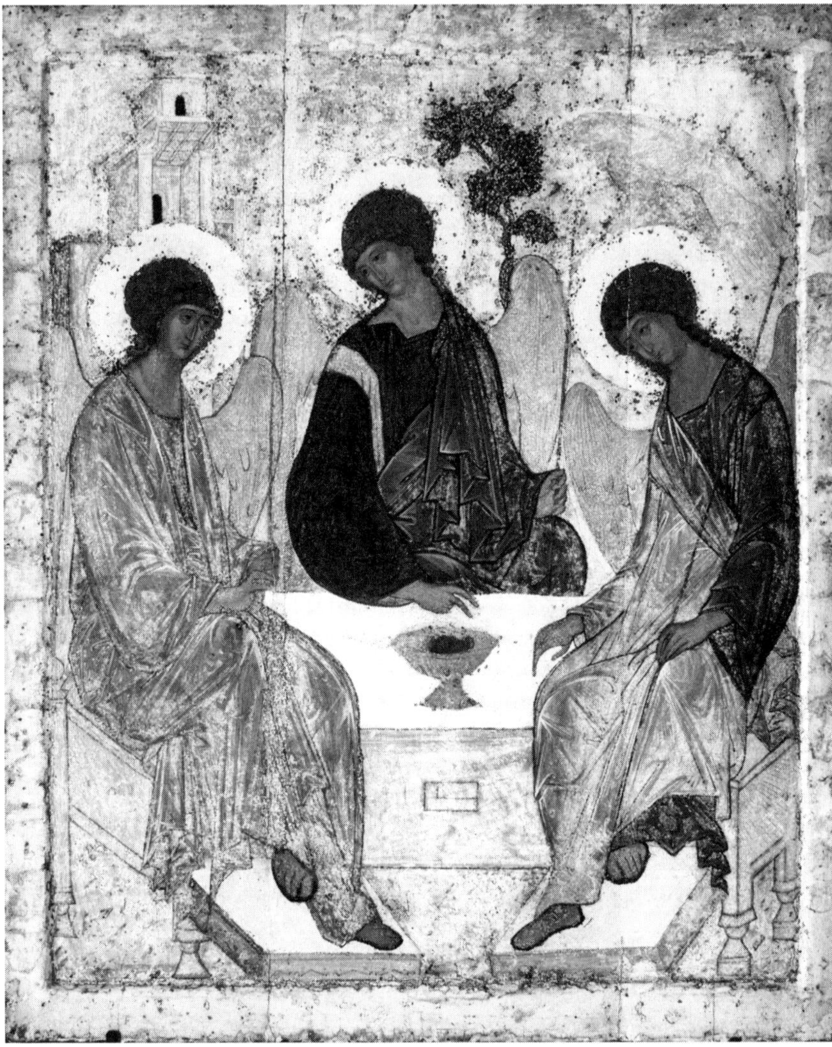

Fig. 5.1 Andrey Rublev, *Holy Trinity,* 1411 (?), tempera on lime board, canvas,State Tretyakov Gallery, Moscow

is characterized by a certain ambiguity of *pictorial space* and, by implication, of *pictorial time.*

An initial problem that faces us is that Rublev's icon is put forward here as an illustration of icon art, which has been defined by an adherence to the Byzantine and Byzantining style. At the same time, there has been a persistent tendency in Russian critical literature to see the *Trinity* as a primary exemplar of a Russian national art, having emerged in its own right parallel to the emergence of Russian national identity. It should not be forgotten that the supposed date of the production of the icon is 1411, when the struggle against the Tartar domination was at its height.[3] In a famous essay, entitled 'Rosiia v ee ikone' (Russia in Her Icons) (1917), Evgeny Trubetskoy (1863–1920) regards Rublev's art as having "expressed the inner history of the Russian religious and national self-consciousness".[4] Florensky, in a similar vein, speaks of Rublev's *Trinity* as a "symbol of the Russian spirit",[5] while Nikolay Tarabukin (1889–1956) claims that the Russian iconographer "liberated himself from Byzantine influences and developed an original (*samobitnii*) style".[6] Another influential critic Nikolay Shchekotov (1884–1945) maintains that Rublev's lyricism is "lacking in hierarchical Byzantine art".[7] The supposed six hundredth anniversary of Rublev's birth in 1960 re-sounds this theme from the inter-war period by stressing the interpretation of the *Trinity* as "deeply national" and "true to the deep national artistic traditions".[8] Andrei Tarkovski's film about the life of Rublev (1966), too, endorses such a nationalistic interpretation. While the rest of the film is in black and white, Rublev's image stands starkly out in the magnificence of its colours and becomes a symbol of the Russian nation in its struggle against the oppressor. The greatness of Rublev lies, it is suggested, in overcoming the Byzantine heritage.[9] The student of Theophanes the Greek – the iconographer, who had come from Constantinople and transplanted Byzantine painting on Russian soil[10] – goes beyond his master and creates a quintessentially Russian image.

On the other hand, however, alongside the "nationalistic" interpretation there has been an articulate "Byzantinist" view which has insisted on the importance of the Byzantine tradition for Rublev's work.[11] For example, scholarly interest

---

[3] The Russian principalities ended formally paying tribute to the Mongols in 1480.

[4] Evgeny Trubetskoy, 'Rossiia v ee ikone' (Russia in Her Icons) (1917) in his *Umozrenie v kraskakh: tri ocherka o russkoi ikone* (Contemplation in Colours. Three Essays on the Russian Icon), (Paris, 1965), pp. 117–18; my translation.

[5] Viktor Bychkov, *The Aesthetic Face of Being: Art in the Theology of Pavel Florensky*, (Crestwood, New York, 1993), p. 37.

[6] The text by Tarabukin was written in 1920–23. Quote from the manuscript from the Tarabukin Archive in the National Library in Moscow cited in Vzdronov, (ed.), *Troitsa Andreia Rubleva*, my translation.

[7] Shchekotov's text was written in 1923. Quote from Vzdronov, (ed.), *Troitsa Andreia Rubleva*, my translation.

[8] Lindsey Hughes, 'Monuments and Identity' in Simon Franklin and Emma Widdis, (eds), *National Identity in Russian Culture*, (Cambridge, 2004), p. 186.

[9] See Andrey Tarkovskii, *Andrei Rublev* (London, 1991).

[10] On Theophanes the Greek, see Viktor Lazarev's well-known *Feofan Grek i ego shkola* (Theophanes the Greek and His School), (Moscow, 1961, German translation from 1968).

[11] Among the earliest works, see Nikolay Punin, *Apollon*, 2 (1915):19–20 and Pavel Muratov, section on Trinity icon in *L'ancienne peinture russe* (Roma, Praha, 1925), pp. 112–13.

has been attracted by the "classicizing" tendencies of Rublev's image, which the Russian iconographer accessed via Byzantine art. Mikhail Alpatov (1902–1986) devoted a number of works to this topic,[12] which had been already discussed by Boris Purishev (b.1903).[13] The great Byzantine scholar Victor Lazarev (1897–1976) proposes that the elegant types of the figures, the motif of the slightly bent head all come from Palaeologan, Constantinopolitan art,[14] while Natalia Demina convincingly suggests that Rublev might have been directly influenced by the Vladimir Mother of God, a Byzantine image of the late eleventh century, which he could have seen both in Moscow and in Vladimir.[15]

Whether Rublev's *Trinity* belongs to a Russian national tradition or to a broader Byzantining, Eastern Orthodox tradition depends on the way we understand the terms of the statement. In the long run, it seems that the controversy is greatly and artificially exaggerated, especially if we accept the view that identity at the time was defined not so much on an ethnic or racial basis but on the basis of Orthodoxy.[16] In Rublev's generation the first reaction of Basil I (1389–1425) at the news of the advance of the feared Timur (Tumbarlane the Great) towards Moscow in 1395 was to send for the sacred icon of Vladimir, which becomes a banner of Orthodoxy. On the day that the icon reached Moscow it become known that Timur had passed on elsewhere.[17] This turn of events was, of course, interpreted accordingly – the image, an accepted *palladium*, believed to have been painted by St. Luke,[18] had saved the city. If Rublev's icon could be seen as expressing identity it did so, I believe, not by overcoming the Byzantining artistic tradition but by consciously identifying with it.

The element of innovation on part of Rublev's is not denied by any means, but it is seen not so much on the level of style, but rather in terms of iconography.

---

12 See, for example, Mikhail Alpatov, 'Klassicheskaia osnova iskusstvo Rubleva' (The Classical Basis of Rublev's Art) in his *Etjudi po istorii russkogo iskusstva*, (Essays on the History of Russian Art) (Moscow, 1963), pp. 112–13, 115–117.

13 Boris Purishev, 'Andrey Rublev i obshchie vaprosi razvitie drevnerusskogo isskustvo 14–17 vv' (Andrey Rublev and Some General Questions of the Development of Ancient Russian Art, 14th – 17th Centuries) in Boris Mikhailovski and Boris Purishev, *Ocherki istorii drevnerusskoi monumental'noi zhivopisi*, (Notes on the History of Ancient Russian Monumental Painting) (Moscow and Leningrad, 1941), pp. 16–19.

14 Viktor Lasareff, 'La Trinité d'André Roublev', *Gazette des Beaux-Arts*, (1959) : 289–300.

15 Natalia Demina, *Troitsa Andreiia Rubleva*, (The Trinity of Andrey Rublev) (Moscow, 1963).

16 On the role of Christianity in establishing a sense of a nation, see Simon Franklin, 'Identity and Religion' in Franklin and Widdis, (eds.), *National Identity in Russian Culture*, pp. 95–116.

17 Bernard Pares, *A History of Russia*, (New York, 1965), p. 85.

18 The status of Vladimir icon is, thus, analogous to the Constantinopolitan *Hodigitria*, which we mentioned.

Rublev treats of a fairly common subject in medieval painting, but at the same time he combines elements of well-known iconographic formulae to a completely new effect. The subject is drawn from the Old Testament and refers to the episode known as Abraham's Hospitality from the Genesis 18:1–8. Abraham and his wife Sarah are visited by three angels, whom they welcome in their home. An interpretation, in the light of the New Testament, holds that the three angels stood for the Holy Trinity. In the early images of the representation of the *Hospitality* the figure of Abraham, with or without Sarah, is usually present (for example, the fourth century fresco in the catacomb on the via Latina in Rome, the fifth cnetury mosaic in Santa Maria Maggiore in Rome, the sixth century mosaic in San Vitale, Ravenna). This remains a common composition in later images, too, but it co-exists with a tendency towards abstraction, to which Rublev's image belongs, whereby the figures of Abraham and Sarah, as well as all secondary details are discarded. In the Russian icon, there is nothing left of the Old Testament story – Abraham and Sarah are missing, as are all indications of a concrete household, of a concrete place. There are no other objects on the table except the chalice with its obvious implication of the Eucharistic sacrifice. Thus, the table itself is transformed into an altar. The lack of narrative elements contributes to an ambiguity of space (there is no indication of location) and time (there is no reference to a time frame of the event). It is, at the same time, interesting to consider that while taking a further step in the direction of abstraction Rublev does not opt, for example, for an altogether symbolic representation of the Trinity (as, for instance, the seventh century mosaic in Nicaea)[19] but stays with the iconographical type of the *Hospitality*.

Three winged figures of almost identical appearance are seated around a table. In other words, there are elements from two distinct iconographical traditions – the circular composition (the angels are arranged in a circle around the table as in Fig. 5.2, an image Rublev must have been well familiar with) and the one following the principle of isocephaly (i.e., the heads of the figures are on the same line, as in a twelfth century mosaic in Monreale in Sicily). With the circular arrangement the angel in the middle is usually represented as bigger in size, which is not the case with the Russian icon. The equal size of figures in Rublev's work clearly derives from the principle of isocephaly.

Not only is it hard to determine whether the figures are male or female, but if we accept them to represent the Trinity it is confusing which figure stands for which person of the Trinity. The latter question has excited a great deal of controversy – learned arguments as to the symbolism of the colours of the garments, of the attitudes towards the other two figures, etc., have been advanced. The most widely held view is Viktor Lazarev's in his book on Rublev, in which the author holds that the central figure is Christ, to his left is God the Father, while to his right is the

19 This image was destroyed in 1922. For details, see Paul Underwood, 'The Evidence Restoration in the Sanctuary Mosaics of the Dormition at Nicaea', *Dumbarton Oaks Papers*, 13 (1959): 235–43.

Holy Ghost.[20] Leonid Ouspensky, after clearly saying that "the icon, by no means, attempts to represent concretely the faces of the Trinity",[21] goes on to side with Lazarev. On the other hand, Demina and Vzdronov[22] believe that the angel in the centre is the Father, while the one to the left is Christ. Most recently, Alexander Voloshinov in his article in *Leonardo* points out the idleness of the question "who is who?" in the icon. Citing G. Pomerantz, he says that if we try to see a difference between the figures of the angels "we are sure to turn the *Trinity* into 'three goats', as Maister Ekhart said".[23] Ironically, Voloshinov, too, feels bounds to take sides on this issue and believes that Christ holds the central position.

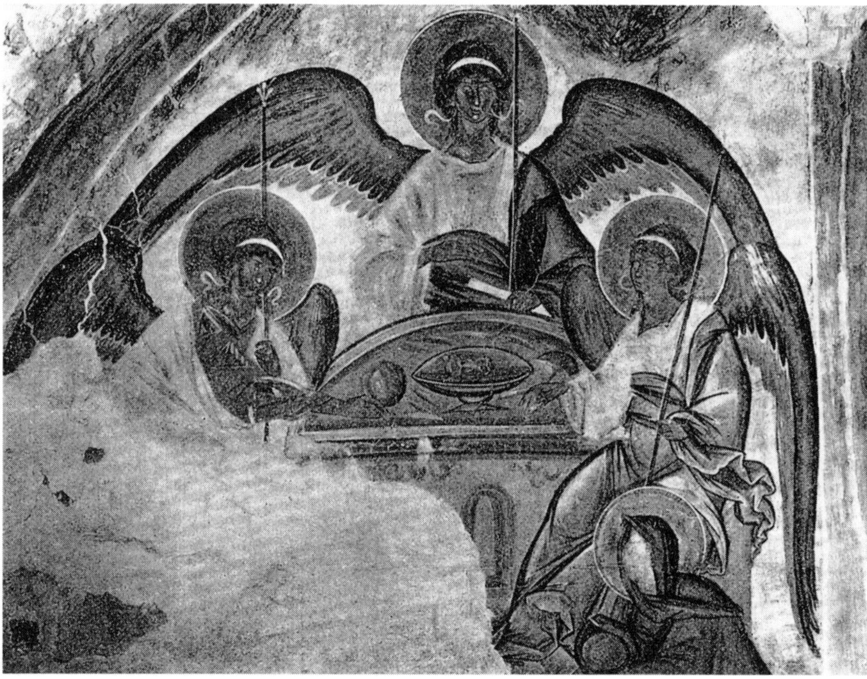

Fig. 5.2 Theophanes the Greek, *The Holy Family*, 1378, fresco, Cathedral of the Transfiguration, Novgorod, Russia

It can be noticed that there are persistent contradictions in Russian writings between the claim that the three figures in Rublev's icon are identical and attempts

20 Viktor Lazarev, *Andrei Rublev*, (Moscow, 1960).

21 Cited in Vzdronov, (ed.), *Troitsa Andreia Rubleva*, p. 95.

22 Gerol'd Vzronov, 'Preface' (in Russian and in English) in Vzdrovov, (ed.), *Troitsa Andreia Rubleva*, p. 12.

23 Alexander Voloshinov, 'The Old Testament Trinity of Andrey Rublyov: Geometry and Philosophy', *Leonardo*, 32/2 (1999): 107.

to determine which angel stands for which person of the Trinity. The whole debate on "who is who", however, seems to be completely misguided as, I believe, it leads us away from Rublev's main intention. Instead I suggest that the three figures are actually *one figure seen simultaneously from different points of view.* It is exactly in this that Rublev diverges from all previous traditions – the figures are almost identical as in paintings using the isocephaly principle, but they are not represented frontally as in this tradition but from different aspects as this happens with the figures in circular compositions. The image could be interpreted as constructed on the principle of simultaneity that was described in last chapter. The difference with the images, referred to in Chapter 2, however, is that that the various aspects of the figure are not synthesized. One of the results is that the figures in Rublev's icon do not look "distorted" to a modern viewer, as it is exactly the process of synthesis, as we saw, that brings about "distortions". Therefore, for instance, the faces of Rublev's figures do not at all resemble the facial type we mentioned – the triangularly-shaped face with disproportionately wide forehead. "Reverse perspective" in the sense we have been using the term, however, still underlies these figures. It is Rublev's deep understanding of "reverse perspective" that allowed him, in a manner, to play with it. The Russian painter puts the principle of simultaneity to a new use that opens new possibilities. Instead of synthesizing the various aspects of the figure, he represents three views of the figure in a manner that recalls "continuous narrative", but differs from it, as there is no time framework (i.e., the Trinity exists outside time, so we cannot expect the representation to mean one figure at three different moments of time). Rublev's approach to the formidable task of painting the Trinity is innovative, but it is of the kind that is steeped in a thorough knowledge of Byzantine artistic principles.

The ensemble of the table/altar with the chalice on it is much closer to the standard use of "reverse perspective" – the chalice is even pushed towards the outer end of the table, a feature, noticed by Zhegin. Compositionally this works very well with the circular arrangements of the figures – we can draw an imaginary line that follows the arch, formed by the three haloes, then goes down the back of each of the side figures, follows the line of the outer arm and continues into the lower part of the chalice. It is only after we "read" the image that the chalice can be placed in the centre of the table. It has a full right to this central position, as, in a way, it is a repetition of the Trinity. After the consecration, it contains Christ, who is God, who is Spirit – in essence (though not in the accidents of bread and wine). The chalice is placed on a table in the form of a trapezoid, a shape which is repeated in the podium. The whole complex, as well as the lack of other objects, serves to emphasize the chalice. Following the guide for reading such images, provided by Florensky and especially Zhegin, we realize that the trapezoid form is the result of the unfolding of the surfaces of a regular square form, of which only the rear parts are lost in the representation. This process of unfolding, evident with the chairs as well, denotes recession without, however, any recourse to optical illusion.

With the pictorial means at his disposal, Rublev departed a long way from the subject of *Abraham's Hospitality*. His icon may well be the most effective visual intuition we possess of the doctrine of the Trinity. In other words, the icon is an example of "theology through art" in a genuine sense of the word and it is, moreover, concerned with nothing less than the "absolute paradox" of Christianity. The Russian iconographer tackles the problem he set himself in a brilliant way but he does that by employing artistic techniques which are very much in the spirit of the Byzantine pictorial tradition and are part of what have been accepted as the basic principles of icon art. This reading, while it might appear forced to a viewer accustomed to linear-perspectival images would have been much more natural to a medieval audience. The reason for the distinction in perception is the different concept of pictorial space. As we saw, iconic space is not systematic and spatial clues are very flexible. In simple words, on the level of spatial clues the representation of three figures does not necessarily imply three figures. In a mathematically constructed painting, unless we have further iconographical information, the number of figures in the painting would be interpreted as immediately seen. In an icon, by contrast, things are not as straight-forward due to the ambiguity of space, and by consequence, the ambiguity of pictorial time.

This seems to me the main import of the image and I think that this interpretation was not only intended by Rublev, but also it was not lost on his audience. It is telling that the Moscow Council of the Hundred Chapters (1551), the so-called *Stoglav* conducted under Ivan the Terrible, pronounced the image under our attention a prototype in its own right. Iconographers were to paint the Trinity from this point on "from ancient models, as painted by Greek painters and as painted by Rublev". Tellingly, the text shows an awareness of a continuity of tradition. *The Tale of the Holy Icon Painters* from second half of the seventeenth century repeats this prescript – icons should be painted "as Andrei Rublev painted them, and not wholly out of one's imagination".[24]

It is worth noticing that it was highly exceptional to accept as prototype, which later iconographers should follow, an image that did not seem to have claims to being *archeiropoietoi* – or in Russian *nerukotvornii* (not made by human hands) – or possess any other miraculous origin. It is on this background that Florensky's discussion of iconography and of Rublev's image in particular is to be understood. According to Florensky, only the saints can be iconographers in the most direct sense of the word.[25] The holy fathers, "with their spiritual experience directed the

---

[24] Cited in Oleg Tarasov, *Icon and Devotion: Sacred Spaces in Imperial Russia*, tr. and ed. R. Milner-Gullard, (London, 2002, first in Russian in 1995), p. 170. On the status of Rublev's works in the medieval period, see Pierre Gonneau, *A l'arbe de la Russie muscovite: Serge de Radonège et André Roublev: Légendes et images (XIVe-XVIIe siècles*, (Institut d'Etudes Slaves, 2007).

[25] Pavel Florensky,'Ikonostas' (Iconostasis) in Pavel Florensky, *Khristianstvo i kul'tura* (Christianity and Culture), (ed.) Alexander Filolenko, (Moscow, 2001), pp. 547 and 563; Pavel Florensky, *Iconostasis*, (Crestwood, New York, 2000), p. 68.

hands of the iconographers who were sufficiently experienced in technique to be able to embody heavenly visions and sufficiently educated to be sensitive to the suggestions of a blessed mentor".[26] The saints are the witnesses of the Biblical message, while iconographers become "witnesses of the Witnesses", "giving us the images, *eicon* of [their] visions"[27]. Thus, "in their artistic form, icons witness directly and visually the reality of this form – they speak, but with lines and colours".[28] The author refers to the text from Nicaea II that stated that "the painter's domain is limited to his art, whereas the disposition (structure, composition, even more – artistic form in general) manifestly pertains to the Holy Fathers"[29]. In the same vein in 1551, the Muscovite Council of One Hundred Chapters, often referred to by Florensky, puts forward the position that "the archbishops and bishops in every city and village, and in every monastery under their care, shall personally examine every master of art, both his life and his art"[30]. Rublev is, in this sense, the vessel of St. Sergius of Radonezd, who is the true creator of the image and it is this that guarantees the *presence* of the prototype in the image. There is no intention to diminish the role of Rublev's artistic genius. Rather, the value of the image is immensely enhanced by shifting the emphasis from subjective creativity to a more objective plain. As Florensky maintains, "there exists the icon of the Holy Trinity by St. Andrei Rublev; therefore, God exists".[31] In other words, the image bears witness to the witness of St. Sergius of the Trinity. Thus, the Trinity icon is seen as a revealed image, i.e., in image revealed in a dream or a vision, and thus it acquires a miraculous status. Rublev, himself a monk at the Trinity monastery founded by St. Sergius, "the worshipper of the Trinity", gives expression in his icon of the saint's mystic vision of the Holy Trinity. St. Sergius's ideal for the transformation of man and the world in the image of the Trinity and the inner unity of all being in God[32] becomes the underlying theme of Rublev's icon.

In a certain sense, the conceptual paradox of the doctrine is solved through the icon, as the image insinuated the doctrine of the Trinity more effectively than any theology could do. On the level of the image, according to the laws of pictorial and no longer conceptual thought, the tension behind "three and yet one" is overcome.

While the aesthetic impact of the image could not be taken as an explanation for its theological status as prototype, this does not exclude that aesthetic beauty

---

26 Ibid., p. 563.

27 Ibid., p. 547; ibid., p. 68.

28 Ibid; my translation; English translation on p. 68 incorrectly renders "khudozestvennoi formoi" which means "artistic form" as "pattern".

29 Ibid, p. 546; English translation by C. Mango (Cyril Mango, *The Art of the Byzantine Empire 312–1453* (Englewood Cliffs, 1972), p. 172); the text in brackets is Florensky's in my translation.

30 Florensky, *Khristianstvo i kul'tura* (Christianity and Culture), p. 569.

31 Ibid., p. 547; Ibid., p. 68.

32 See Trubetskoy, 'Umozrenie v kraskakh' (Contemplation in Colours), p. 20.

has been a factor in the appreciation of images not only by its modern public but in a medieval context as well. The *Stoglav* explicitly places importance on the aesthetic dimension of icon-painting: "And even though a person should lead a good spiritual life, if he is incapable of designing holy icons beautifully, such a person should not be permitted to paint holy icons, but should earn his nourishment by some other craft".[33] The passage is interesting for making explicit an idea that was by no means absent with Byzantine theology of the image but was rarely expressed in such obvious terms. In Russia, on the other hand, as has been frequently remarked, what can be taken as aesthetic factors affected the very conversion to Christianity. The much cited passage from *The Russian Primary Chronicle* recounts the reaction of ambassadors of Prince Vladimir of Kiev, when they saw Hagia Sophia in Constantinople: "We knew not whether we were in heaven or on earth, for surely there is no such splendour or beauty anywhere upon earth. We cannot describe it to you. We only know that God dwell there among men. For we cannot forget that beauty".[34] The context is the missions that Vladimir had sent to investigate the advantages and disadvantages for a conversion of the Russian people to one of the big religions of the region – Islam, Catholicism and Orthodoxy. We can see from the *Chronicle* that the aesthetic dimension was not a separate issue, but was accepted as part of the "package" and therefore, the modern isolation of aesthetics from theology would not hold true in this context.

In this sense, the debate on Rublev's work as being a "genuine (*nastoiashchaia*) icon" (Alpatov) or, conversely, not an icon at all (Shchekov) is another example of applying mechanically modern categories to medieval phenomena. The connotation of the term "icon" here comes down to a "religious image", the meaning of which can be accessed through theology. In a highly politicized context, an opposition was drawn between the work as a "cult object" versus the work as an "aesthetical object". The new meaning that the *Holy Trinity* would acquire when moved from the monastery to the gallery was addressed, for instance, by Nikolay Kuz'min[35] and Feodor Odinokov.[36] Actually, as we saw, the theological view needs to be combined with visual studies in order to address the problem of how a visual image can intuit a theological dogma, analyzable in conceptual terms.

Rublev's icon has provided an inexhaustible source of inspiration for new interpretations, partly because it can be seen on various levels of meaning. In this text I have tried to approach it through the categories and notions that were evolved through this book.

---

[33] Cited in Tarasov, *Icon and Devotion: Sacred Spaces in Imperial Russia*, p. 179.

[34] *The Russian Primary Chronicle*, tr. and ed. Samuel H. Cross and Olgerd Sherbowitz-Wetzor, (Cambridge, MA, 1953), p. 198.

[35] Nikolay Kuz'min, 'Andrey Rublev', *Novyi mir*, 10 (1960): 206.

[36] Feodor Odinokov, 'Obretenie Andreia Rubleva', *Nauka i religiia*, 10 (1971): 86.

Glossary of Terms

Icon: The term "icon" is most commonly used with respect to religious images done in the "Byzantine and Byzantining style".[1] "Byzantining" comprehends post-Byzantine art, as well as art, done in the Byzantine manner outside of Byzantium even before the fall of Constantinople under the Ottomans. The former is an aspect of what the Romanian scholar Nicolai Iorga has called "Byzance après Byzance"[2] and refers to the continued production of art, following the Byzantine tradition after 1453. The latter refers to the adaptation of Byzantine forms during the Middle Ages within the territories comprising the "Byzantine Commonwealth" (the Balkans, Romania, and Russia),[3] but also outside the Eastern Orthodox world, as in the *maniera greca*, which was at its height in thirteenth century Italian art.

The term itself comes from the Greek *eikon* which means "likeness, image, picture".[4] Until about the fifth century it had a more general usage, referring to any kind of portrait, whether of a saint or a common mortal. From the sixth century onwards the word *graphis* was introduced to describe a secular portrait, while *eikon* was reserved for religious paintings only.

What is of interest for our purposes is the implication of "likeness". For ancient consciousness an icon is an image, pictorial or any other, including mental ones, which presupposes a "likeness" between the image and the prototype (see "prototype"). In a Christian context the connection between "image" and "likeness" is made in the Book of Genesis 1. 26 and 27: "Let us make man to our image and likeness [...] And God created man to His own image; to the image of God He created him". The "likeness", however, does not imply naturalistic resemblance,[5] but a recognizable resemblance to a "form" (see "form") that is

---

1 Erwin Panofsky, 'History of the Theory of Human Proportion as a Reflection of the History of Styles' in his *Meaning in the Visual Arts*, (Harmondsworth, 1970), p. 109.

2 Nicholai Iorga, *Byzance après Byzance: Continuation de l'Histoire de la vie byzantine* (Bucharest, 1935).

3 Dmitri Obolensky, *The Byzantine Commonwealth*, (London, 1971).

4 Geoffrey W. H. Lampe, (ed.), A *Patristic Greek Lexicon*, (Oxford, 1961), pp. 410-16. See also Gerhart Ladner, 'The Concept of the Image in the Greek Fathers and the Byzantine Iconoclastic Controversy', *Dumbarton Oaks Papers* 7, (1953): 3-34. Ladner's is probably the most useful concise account of the development of the concept of the image from patristic thought on the background of Hellenistic notions to the Byzantine iconophiles.

5 That the likeness of the image does not claim illusionistic resemblance, but is "something spiritual-intellectual" is "the almost unanimous opinion of the Fathers of the

authorized by tradition and accepted by the Church to carry the presence of the prototype in the image. One of the aims of this book has been to define the nature of this presence, which I call "partial real presence" to indicate the "conception of the image as a blend of like and unlike, same and other"[6] (see "presence") and which gives grounds for defining the icon as symbol, in a specific sense of the term (see "symbol"). Another objective is to define "reverse perspective" (see "reverse perspective) as an aspect of the form which contains presence and in this way becomes an element in the very definition of the icon.

Prototype: Prototype comes from the Greek *prototypos*, which means "first-formed", "first, primary" as well as "original, prototypal, i.e., from which subsequent models derive".[7] In the sense of iconic images, a prototype is the original icon of a particular subject. Thus, a certain analogy could be drawn with the distinction between original and copy. A prototype is an image that has been acknowledged as original by the Church and hence can serve as a model to a series of copies.

Commonly, a prototype is believed to have divine or at least exceptional origin. The paradigm is supplied by the Logos or God, who became flesh that is the prototype for the Christ image. There are instances, however, of obviously man-made prototypes, as Andrey Rublev's icon of *The Holy Trinity*, which was confirmed as a new canonical form by several Russian Councils (including The Councils of the Hundred Chapters in the sixteenth century, which dealt specifically with icon-painting). Important repercussions for aesthetics can be drawn from the potentiality of a new image to become a prototype and some attention has been paid to this issue by Florensky in the *Iconostasis*, as well as by the contemporary Russian scholar Victor Bychkov in his *Vizantiiskaia estetika* (Byzantine Aesthetics).

Symbol: One of the ways in which the icon is discussed in my work is in its role as a symbol. Among the several main definitions of the symbol, the one that is of interest here posits an ontological identity between symbol and that which it symbolizes, or in other words, between icon and prototype. Thus, the image becomes a "special kind of symbol, which presents to our transitory world that which it depicts".[8]

Church" (with the exception of Mathodius) since the third century, according to Ladner (Ladner, 'The Concept of the Image in the Greek Fathers and the Byzantine Iconoclastic Controversy', 11). This view is connected to the interpretation of *mimesis* in patristic thought.

6 Ladner, 'The Concept of the Image in the Greek Fathers and the Byzantine Iconoclastic Controversy', 12.

7 Lampe, *A Patristic Greek Lexicon*.

8 Viktor Bychkov, 'Icon' in M. Kelly, (ed.), *Encyclopaedia of Aesthetics*, (New York and Oxford, 1998), vol. 2, p. 449.

There are at least two traditions of thought which subscribe to the basic principle of this understanding and which are of immediate interest for our purposes – Byzantine theology of the icon with an antecedent in Pseudo-Dionysius on the one hand, and German romantic theories of the symbol on the other. I have focused specifically on aspects of these two trends as sources to Florensky's understanding on the problem.

Reverse perspective: The term "reverse" – or "inverse" – perspective is the English translation of the German "*umgekehrten Perspektive*", popularized by Oscar Wulff in his article "Die umgekehrten Perspektive und die Niedersichte" (1907), where the author proposes that the spatial organization typical of Byzantine images turns around the rules of linear perspective. Florensky borrows Wulff's term in his essay "Reverse Perspective" (read in 1920, published in 1967) and accepts the German's author's main conception which he develops further in the same direction, i.e., the pictorial space of icons reverses the laws governing mathematically constructed space. The main writings on "reverse perspective" were produced by Russian authors and they were, almost without exception, heavily indebted to Florensky and, on the whole, followed in the same pathway. This idea has also entered Western scholarship mainly via Wulff, with or without knowledge of Russian works on the subject, but there it has never been studied systematically.

I have used the term "reverse perspective" in inverted commas to emphasize my belief that the commonly accepted view is conceptually flawed and does not explain the principle at the heart of the phenomenon under our attention. Rather, I accept a definition of "reverse perspective" as a principle of spatial organization of the icon, which is characterized by "simultaneous planes", a feature first noticed by Florensky (Florensky's own term is "supplementary" or "additional planes"). The present understanding of "reverse perspective" refers to the simultaneous representation of different aspects of the image, whether they can be seen at the same time from a fixed position or not, on the same picture plane. It is noteworthy that the phenomenon of "simultaneous planes" is mentioned by all other major Russian writers on "reverse perspective" (Lev Zhegin, Boris Uspensky), but goes largely unnoticed in Western scholarship. Neither Florensky nor anyone after him, however, pursued the logical conclusions of this observation nor identified the contradiction it presented to the traditional definition of "reverse perspective" (see "reverse perspective"). My thesis is that "reverse perspective" understood in this manner is underlined by a certain theological concept of time, namely the doctrine of timeless eternity (see "concept of simultaneity").

Reverse time: This is a term coined by Florensky to describe the time flow in dreams and art – more particularly icon art – and it represents the only consistent attempt that I am aware of, of proposing a conception, describing the temporal dimension of the icon. The term, in my opinion, was meant as a correspondent to "reverse perspective" and it discloses the same approach as used in Russian theory on "reverse perspective". In this case, linear time is being reversed in dreams and art. The misconception here is that it is suggested that the reversal is in terms of direction, while, in fact, there is a reversal of the speed of time, as well. In this sense, we are talking of a fundamentally different phenomenon, not just of an "opposite", so a better term is desirable.

Presence: There is a line of thought, going back to pagan and Christian Neoplatonism and traced in this book through Byzantine theology of the image of the eighth and ninth centuries, the German romantics and Russian religious philosophy as represented by Pavel Florensky, which proposes that the prototype is present in its image. All the various formulations on the nature of this presence come down to what I will call partial real presence. The idea is that the image contains the presence of the prototype in some, but not all, aspects. A full, in contrast to partial, presence would amount to magical beliefs and this is frequently the manner in which the lay believer interprets the holy image. With Byzantine Iconophile writers, though, an elaborate theory was worked out according to which the prototype was present in terms of Aristotelian accident, but not in terms of essence. In this sense, the image was *homoios* (similar to) rather than *homoousios* (essentially the same) with the prototype (see *homos*, *homoousios*, *homoios*, *homoioisios*). In other words, there is a reality of presence which, however, is different from the Real Presence in the Eucharist in which there is a change of substance (hence, the Catholic "transubstantiation"), while the accidents remain the same. The analogy, however, between the mystery of the Eucharist and the prototype-image relationship has been made by some authors, both Byzantine and contemporary, and it is useful as it clarifies the nature of the relationship in both cases as partial and as explained through Aristotelian terminology.

The concept of "partial real presence" admittedly rests upon a paradox in the sense that Kierkegaard uses this term in various of his writings.[9] Christ is the absolute paradox exactly, in that being God he became at the same time a particular man, which is "the greatest possible, infinitely qualitative distance from being God" and so, as Kierkegaard says, "the most profound incognito".[10] Christ, however, is both the image of God (Corinthians 4,3; Colossians 1,15, etc.) and the prototype of his icon. As Gerhart Ladner says, "the Divine Logos

[9] For instance, in *Fear and Trembling* (1843), *Repetition* (1843), *Philosophical Fragments* (1844), *Concluding Unscientific Postscript* (1846), *Practice in Christianity* (1850).

[10] Søren Kierkegaard, *Practice in Christianity*, (Princeton, 1991, first in Danish in 1850), p. 127.

Himself becomes the image of God, and even the images of such an image participate in its divine character".[11]

The logical contradiction at the heart of the icon is a theme, running throughout the Byzantine theology of the image and it is in harmony with the spirit of Byzantine theology in general, which is based on antinomies, i.e., propositions which are mutually exclusive in formal logic.[12]

Homoousios – Homoiousios: "*Homoousios*" means "of the same essence" (from *homos* which is "identical with", "essentially the same"), while "homoiousios" can be translated as "like" (like but not the same) from *homoios*, which is "similar to", "comparable to"). The distinction should be kept in mind to understand the frequently complicated Iconoclastic and Iconophile arguments that were put forward in Byzantium. The Iconoclasts tended to understand – or rather, misunderstand – their adversaries' position as suggesting that the icon was "*homoousios*" to the prototype, while the Iconophiles themselves insisted on similarity, rather than identity.

Form (*morphe* and *character*): Form is a highly ambiguous term. *Morphe* can be used in the pair "form and matter" and in this case it can mean "soul", as Aristotle uses the term in *De Anima*, i.e., the principle that brings matter into being. *Morphe* also has the connotation of "shape, external appearance", and in this sense, it can be used as a synonym to "accident". *Character* has this latter meaning of *morphe.* As *character* also means "seal" and "imprint" when it is used in the sense of "form" it keeps this additional connotation of "seal" and "imprint". It is important to realize that when we define the relationship between "icon" and "prototype" as shared *form*, all these connotations are at play. What is shared is external appearance/*morphe*, but also what is shared has been transmitted in the manner that a seal makes an imprint. Hence arose the talk about a sort of identity and a sort of similarity.

Concept of simultaneity: There are several main connotations of "simultaneity" that have been referred to in this book. According to the first, theological meaning, simultaneity follows from the doctrine of a timelessly eternal God. As God is not subject of the laws of time, he exists simultaneously, i.e., to him present, past, and future exist all at once. For our purposes, this concept is of importance as it implies a simultaneous vision, i.e., one that perceives objects not successively in space, but simultaneously, all at once. For this perception there is no point of view, since it is outside the laws of space as well. Further, in a more

---

[11] Ladner, 'The Concept of the Image in the Greek Fathers and the Byzantine Iconoclastic Controversy', 5.

[12] John Meyendorff, 'Conclusion: Antinomies' in *Byzantine Theology: Historical Trends and Doctrinal Themes* (New York, 1974), pp. 224-5. On the antinomical character of the icon see also Bychkov, 'Icon', p. 450.

modern context which can be understood outside the theological background, simultaneous experience was an important part of the modernist experiments in art, literature and music at the beginning of the twentieth century. Florensky's notion of "supplementary planes" (see "simultaneous planes") is indebted to both of these trends of thought. Finally, from a scientific point of view, Einstein has proved the impossibility of absolute simultaneity.[13]

[13] See Albert Einstein, 'On the Electrodynamics of Moving Bodies' in Hendrik Lorentz et al., *The Principle of Relativity: A Collection of Original Memoirs on the Special and General Theory of Relativity*, (New York, 1952, first in English translation in 1923).

Bibliography

Acheimastou-Potamionau, Myrtali (ed.), *From Byzantium to El Greco: Greek Frescos and Icons* (London, 1987).

Adams, Henry, *The Education of Henry Adams: An Autobiography* (Boston, 1918, first in 1907).

Ades, D., 'Art and Time in the Twentieth Century' in Kristen Lippincott with Umberto Eco, Ernst Gombrich et al., *The Story of Art* (London, 1999), pp. 202–18.

Adorno, Theodor and Max Horkheimer, *Dialectic of Enlightenment* (London, 1986).

Alpatov, Mikhail, 'Klassicheskaia osnova iskusstvo Rubleva' (The Classical Basis of Rublev's Art) in his *Etjudi po istorii russkogo iskusstva*, (Essays on the History of Russian Art) (Moscow, 1963).

Alpers, Svetlana, and Paul Alpers, '*Ut Pictura Noesis?* Criticism in Literary Studies and Art History', *New Literary History*, 3/3 (1972): 437–58.

Andrews, Lew, *Story and Space in Renaissance Art: The Rebirth of Continuous Narrative* (Cambridge, 1999).

Anselm, *Proslogion*, chapter 19 in F. Schmitt, (ed.), *S. Anselmi: Opera omnia* (Edinburgh, 1946).

Antonova, Clemena, '"Beauty Will Save the World": The Revival of Romantic Theories of the Symbol in Pavel Florensky's Works', *Slavonica*, 14/1 (2008): 44–56.

——, 'Re-contextualizing Holy Images in Early Soviet Russia: Florensky's Response to Lenin's Plan for Monumental Propaganda' in *Der Sturm der Bilder/ The Clash of Images*, published for the University of Hamburg, forthcoming.

——, with Martin Kemp, '"Reverse Perspective": Historical Fallacies and an Alternative View' in Michele Emmer (ed.), *The Visual Mind II* (Cambridge, MA, 2005), pp. 524–66.

——, 'Florensky's "Reverse Time" and Bakhtin's "Chronotope" – a Russian Contribution to the Theory of the Visual Arts', *Slovo*, 15/2 (2003): 101–114.

Apollinaire, Guillaume, *The Cubist Painters* (1913) (Harrogate, 2000).

Armstrong, Arthur (ed.), *The Cambridge History of Later Greek and Early Medieval Philosophy* (Cambridge, 1967).

——, 'Eternity, Life and Movement' in *Le Néoplatonisme* (Paris, 1971), pp. 67–74.

Artemidorus Daldianus, *Das Traumbuch* (ed.) K. Blackertz (Munich, 1979).

Asmussen, Hans, *Die Lehre vom Gottesdienst* (Munich, 1937).

Atwell, John, 'Art as Liberation: A Central Theme of Schopenhauer's Philosophy' in Dale Jacquette (ed.), *Schopenhauer, Philosophy, and the Arts* (Cambridge, 1996), pp. 81–107.

Auzépy, Marie-France, 'La destruction de l'icône du Christ de la Chalcé par Leon III : propagande ou réalité?', *Byzantion*, 60 (1990): 445–492.

——, *La vie d'Étienne le Jeune par Étienne le Diacre* (Aldershot, 1997).

Babushinski, Andrei, 'Lineinaia perspektiva v iskusstve i zritel'nom vozpriatii real'nogo prostranstvo' (Linear Perspective in Art and the Visual Perception of Real Space), *Iskusstvo*, 1/1 (1923).

Baggley, John, *Doors of Perception: Icons and Their Spiritual Significance* (London and Oxford, 1987).

Bakhtin, Mikhail, 'Problema teksta v lingvistiki, filologii, i drugikh gumanitarnikh naukakh. Opit filosofskogo analiza' (The Problem of the Text in Linguistics, Philology and the Other Humanities. An Attempt at Philosophical Analysis) in *Estetika slovesnogo tvorchestvo* (The Aesthetics of Linguistic Creativity) (Moscow, 1979).

——, 'Forms of Time and Space in the Novel' (1937–8) in Michael Holquist (ed.), *The Dialogical Imagination: Four Essays by M. M. Bakhtin* (Austin, TX, 1981), pp. 84–258.

——, *Rabelais and His World* (Bloomington, IN, 1984).

Barasch, Moshe, *Icon: Studies in the History of an Idea* (New York and London, 1992).

Barber, Charles, *Figure and Likeness: On the Limits of Representation in Byzantine Iconoclasm* (Princeton and Oxford, 2002).

Barr, James, *Biblical Words for Time* (Naperville, IL, 1962).

Barrow, John D., *Impossibility: The Limits of Science and the Science of Limits* (London, 1998).

Basil the Great, 'On the Hexaemeron' in *Fathers of the Church*, vol. 46 (Washington, 1963).

Batteux, Abbé, 'The Fine Arts Reduced to a Single Principle' (1746) in Susan Feagin and Patrick Maynard (eds), *Aesthetics* (Oxford and New York, 1997), pp. 102–105.

Baxandall, Michael, *Painting and Experience in Fifteenth-century Italy* (Oxford and New York, 1972).

Begbie, Jeremy, *Theology, Music and Time* (Cambridge, 2000).

Belting, Hans, 'An Image and Its Function in Liturgy', *Dumbarton Oaks Papers*, 34–5 (1980–81): 1–17.

——, *Likeness and Presence* (Chicago and London, 1994).

Berdyaev, Nikolay, *The Meaning of History* (London, 1936, reprint 1945).

Bergman, Pär, *Modernolatria et 'Simultaneity': Recherches sur deux tendances dans l'avant-garde littéraire en Italie et en France á la veille de la première guerre mondiale* (Stockholm, 1992).

Bernstein, Jay, *The Fate of Art: Aesthetic Alienation from Kant to Derrida and Adorno* (Cambridge, 1992).

Besançon, Alain, *The Forbidden Image: An Intellectual History of Iconoclasm* (Chicago and London, 2000, first in French in 1994).

Boer, Willem der, 'Greco-Roman Historiography in Its Relation to Biblical and Modern Thinking', *History and Theory*, 7/1 (1968): 72.

Bonnefoy, Yves, 'Time and the Timeless in Quattrocento Painting' in Norman Bryson (ed.), *Calligram: Essays in New Art History from France* (Cambridge, New York, 1988), pp. 8–27.

Brandist, Craig, *The Bakhtin Circle: Philosophy, Culture and Politics* (London and Sterling; Virginia, 2002).

Brandon, Samuel, *Time and Mankind* (London, 1951).

Brown, Peter, *The Cult of Saints: Its Rise and Function in Latin Christianity* (Chicago and London, 1981).

——, *Society and the Holy* (Chicago, 1982).

Bruns, Gerald, *Heidegger's Estrangements: Language, Truth and Poetry in the Later Writings* (New Haven and London, 1989).

Bunim, Miriam, *Space in Medieval Painting and the Forerunners of Perspective* (New York, 1940).

Bychkov, Viktor, *Vizantiiskaia estetika* (Byzantine Aesthetics) (Moscow, 1977).

——, *Russkaia srednevekovnaia estetika XI – XVII stoletia* (Russian Medieval Aesthetics XI – XVII Centuries) (Moscow, 1992).

——, *The Aesthetic Face of Being: Art in the Theology of Pavel Florensky* (Crestwood, New York, 1993).

——, 'Icon' in Michael Kelly (ed.), *Encyclopaedia of Aesthetics*, vol. 2 (New York and Oxford, 1998), pp. 448–50.

Bychkov, Viktor, and I. Kish, 'Vidna li neveroiushtemu Troitza Rubleva?: K voprosu o poniiatii ikonnogo znaka v '*Ikonostase*' P. Florenskogo' (Is Rublev's Trinity Visible to the Non-believer?: On the Question of the Iconic Sign in P. Florensky's *Iconostasis*) in Michael Hagemeister and Nina Kauchtschischwili (eds), *P. A. Florensky i kul'tura ego vremeni* (P.A. Florensky and the Culture of His Time) (Marburg, 1995), pp. 399–411.

Cameron, Averil, 'The Language of Images: the Rise of Icons and Christian Representation' in Diana Wood (ed.), *The Church and the Arts* (Oxford and Cambridge, MA, 1992), pp. 1–43.

——, *The Late Roman World: 384–430 AD* (London, 1993).

Carlson, Maria, *'No Religion Higher than Truth': A History of the Theosophical Movement in Russia, 1875–1922* (Princeton, 1993).

Carroll, Lewis, 'Alice's Adventures in Wonderland' in Lewis Carroll, *The Penguin Complete Lewis Carroll*, (Harmondsworth, 1982).

Cassirer, Ernst, *Kant's Life and Thought* (New Haven and London, 1981).

Christov, Ivan, 'Introduction' (in Bulgarian) to his Bulgarian translation of Pseudo-Dionysius, *The Heavenly Hierarchy and the Ecclesiastical Hierarchy* (Sofia, 2001).

Compton, Susan, 'Malevich and the Fourth Dimension', *Studio International*, vol. 187, April 1974.

——, 'Malevich's Suprematism – the Higher Intuition', *Burlington Magazine*, vol. CXVIII, 1976, 577–85.

Coomaraswamy, Ananda, 'An Early Passage on Eastern Painting', *Eastern Art*, 3 (1931): 218–9.

——, 'Meister Eckhart's Vew on Art' in his *The Transformation of Nature in Art* (Cambridge, MA, 1934), pp. 59–97.

Cooper, Douglas, *The Cubist Epoch* (New York, 1971).

Cooper, J., 'Plato on Sense-Perception and Knowledge (*Theaetetus 184–186*)' in Gail Fine (ed.), *Plato I: Metaphysics and Epistemology* (Oxford, 1999), pp. 356–75.

Cormack, Robin, *Writing in Gold* (London, 1985).

——, 'Byzantine Aphrodiasias. Changing the Symbolic Map of a City', *Proceedings of the Cambridge Philological Association*, 216/36 (1990): 26–41.

——, 'The Icon, Past, Present, and Future' in Roderick Grieson (ed.), *Gates of Mystery: The Art of Holy Russia* (Cambridge, 1994), pp. 322–30.

——, *Painting the Soul: Icons, Death Masks and Shrouds* (London, 1997).

Cornford, Francis (tr. and commentary), *Plato's Cosmology: the Timaeus of Plato* (London, 1937).

Coxon, A.H., *Philosophy of Forms: An Analytical and Historical Commentary on Plato's 'Parmenides' with a New English Translation* (Assen, Netherlands, 1999).

Craig, Edward, *The Mind of God and the Mind of Man* (Oxford, 1987).

Cranston, Sylvia, *HPB: The Extraordinary Life and Influence of Helena Blavatsky, Founder of the Modern Theosophical Society* (New York, 1993).

Crestwell, Keppel, 'The Lawfulness of Painting', *Ars Islamica*, 11–12 (1946): 159–66.

Crowther, Paul, *Critical Aesthetics and Postmodernism* (Oxford, 1993).

——, *Art and Embodiment: From Aesthetics to Self-consciousness* (Oxford, 1993).

——,'The Epistemological Significance of the Moment', *Issues of Contemporary Culture and Aesthetics*, 8 (1998): 31–6.

——, *The Transhistorical Image: Philosophizing Art and Its History* (Cambridge, 2002).

——, *Defining Art, Creating the Canon* (Oxford, 2007).

Cullmann, Oscar, *Christ and Time: The Primitive Christian Conception of Time and History* (London, 1957, first in 1951).

Curtis, James, 'Shestov's Use of Nietzsche in His Interpretation of Tolstoy and Dostoevsky', *Texas Studies in Literature and Language*, 27 (1975).

Dagron, Gilbert, 'Holy Images and Likeness', *Dumbarton Oaks Papers*, 45 (1991): 23–33.

Dante, *Vita Nuova* (Oxford and New York, 1998).

Davis, Stephen, *Logic and the Nature of God* (London, 1983).

Davis, Whitney, *The Canonical Tradition in Ancient Egyptian Art* (Cambridge, 1989).

Demina, Natalia, *Troitsa Andreiia Rubleva*, (The Trinity of Andrey Rublev) (Moscow, 1963).

Dewey, John, *Art as Experience* (London, 1934).

Diffey, T. J., 'Schopenhauer's Account of Aesthetics Experience', *The British Journal of Aesthetics*, 30/2 (1990): 132–42.

Dillenberger, John, *Images and Relics: Theological Perceptions and Visual Images in the Sixteenth Century Europe* (Oxford, 1999).

Dio Cassius, *Dio's Roman History*, vol.64 (London, 1954-1955).

Dobschutz, E. von, *Christusbilder* (Leipzig, 1899).

Doehlemann, Karl, 'Zur Frage der sog.' "umgekehrten Perspektive"', *Repertorium fur Kunstwissenschaft*, 1910.

Dostoevsky, Fyodor, *The Brothers Karamazov*, vol. 1 (Moscow, 1980).

——, *Bratia Karamazovi* (The Brothers Karamazov) in Fyodor Dostoevsky, *Polnoe sobranie sochinenii* (Complete Works), Leningrad, vol. 14.

Doxiades, Euphrosyne, *The Mysterious Fayum Portraits: Faces from Ancient Egypt* (London, 1995).

Dryden, John, *The Essays of John Dryden*, ed. W. Ker (New York, 1961).

Dubisch, Jill, 'In a Different Place' in Ellen Badone (ed.), *Religious Orthodoxy and Popular Faith in European Society* (Princeton, *c.*1990), pp. 113–39.

Eckhart, Meister, *Meister Eckhart*, tr. C. Evans (London, 1924–31).

Eco, Umberto, *Art and Beauty in the Middle Ages* (New Haven and London, 1986, first in Italian in 1959).

Edelstein Ludwig, *The Idea of Progress in Classical Antiquity* (Baltimore, 1967).

Einstein, Albert, 'On the Electrodynamics of Moving Bodies' (1905) in Hendrik Lorentz et al., *The Principle of Relativity: A Collection of Original Memoirs on the Special and the General Theory of Relativity* (New York, 1952, first in English translation in 1923), pp. 35–65.

——, *The Meaning of Relativity* (Princeton, 1923), p. 33.

Elkins, James, *The Poetics of Perspective* (Ithaca and London, 1994).

——, *The Object Stares Back: On the Nature of Seeing* (New York and London, *c.*1996).

——, *On Pictures and the Words That Fail Them* (Cambridge, 1998).

Jas Elsner, 'Image and Iconoclasm in Byzantium', *Art History*, 11/4 (1988): 477–81.

——, *Imperial Rome and Christian Triumph* (Oxford and New York, 1998).

——, 'Between Mimesis and Divine Power: Visuality in the Greco-Roman World' in Robert Nelson (ed.), *Visuality before and beyond the Renaissance* (Cambridge, 2000), pp. 45–70.

Emerson, Caryl (ed.), *Critical Essays on Mikhail Bakhtin* (New York, 1999).

Etkind, Aleksander, *Eros nevozmoznogo: Istoriia psikhoanaliza v Rossii* (Eros of the Impossible: History of Psychoanalysis in Russia) (Moscow, 1994, English translation from 1997).

Evagrius, Scholasticus, *The Ecclesiastical History of Evagrius: With the Scholia* (Amsterdam, London, 1898, reprint 1964).

Fiedler, Konrad, 'Observation on the Nature and History of Architecture' (1878) in Harry Mallgrave and Elephtherios Ikonomou (eds), *Empathy, Form and Space: Problems in German Aesthetics 1873–1893* (Santa Monica, CA, 1994), pp. 125–46.

Fitzgerald, Paul, 'Stump and Kretzmann on Time and Eternity', *Journal of Philosophy*, 82/5 (1985): 260–69.

Florensky, 'Ikonostas' (Iconostasis) in *Pavel Florensky: Khristianstvo i kul'tura* (Pavel Florensky: Christianity and Culture), ed. A. Filonenko (Moscow, 2001); English translation in Florensky, P., *Iconostasis* (Crestwood, New York, 1996, reprint 2000).

——, *Sobranie sochineniia* (Complete Works), (Paris, 1985).

——, *Beyond Vision: Essays on the Perception of Art*, ed. N. Misler (London, 2002).

——, *The Pillar and Ground of Truth*, tr. and annotation by B. Jakim (Princeton, 1997).

——, *Sochineniia. v chetirekh tomakh* (Works in Four Volumes) (Moscow, 1999).

Florovsky, Georges, 'The Worshipping Church' in Orthodox Eastern Church, *The Festal Menaion*, tr. Mother Mary and Archimandrite K. Ware (London, 1969), pp. 21–38.

——, *Collected Works*, vol. 7: *The Eastern Fathers of the Church* (Belmont, MA, 1972–89).

Franklin, Simon, *Writing, Society and Culture in Early Rus, c.950–1300* (Cambridge, 2002).

——, 'Identity and Religion' in Simon Franklin and Emma Widdis (eds.), *National Identity in Russian Culture* (Cambridge, 2004), pp. 95–116.

Freedberg, David, *The Power of Images: Studies in the History and Theory of Response* (Chicago and London, 1989).

Fry, Edward, *Cubism* (London, 1966).

Gandelman, Claude, 'The 'Scanning' of Pictures', *Communication and Cognition*, 19/1 (1986): 3–24.

Gavin, Frank, 'The Eucharist in the East and West' in W. K. Lowther Clarke (ed.), *Liturgy and Worship* (London, New York, 1932), pp. 78–130.

Giakalis, Ambrosios, *Images of the Divine: The Theology of Icons at the Seventh Ecumenical Council* (Leiden, New York, 1994).

Frank, Joseph, 'Spatial Form in Narrative' in Jeffrey Smitten and Ann Daghistany (eds), *Spatial Form in Narrative* (Ithaca, 1981).

Freud, Sigmund, *The Interpretation of Dreams* (New York, 1965, first in 1900).

Gibbons, Tom, 'Cubism and the 'Fourth Dimension' in the Context of Late Nineteenth Century and Early Twentieth Century Revival of Occult Idealism', *Journal of the Warburg and Courtauld Institutes*, 44 (1981): 130–47.

Giovanni, G., 'The Method of Study of Literature in Its Relation to the Other Fine Arts', *Journal of Aesthetics and Art Criticism*, 8 (1949).

Gleize, Albert and Jean Metzinger, *Du Cubisme* (Paris, 1912).

Goethe, *Poetry and Truth: From My Own Life* (London, 1908).

Golding, John, *Cubism: A History and an Analysis* (London, 1968, first in 1959).
Gombrich, Ernst, Lessing (Lecture on a Master Mind)' in *The Proceedings of the British Academy* (London, 1957), pp. 133–56.
——, *Art and Illusion* (London and New York, 1960).
——, 'The Evidence of Images' in Charles Singleton (ed.), *Interpretation: Theory and Practice* (Baltimore, 1969), pp. 35–105.
——, *Aby Warburg: An Intellectual Biography* (Oxford, 1986, first in 1970).
——, *The Image and the Eye: Further Studies in the Psychology of Pictorial Representation* (London, 1982).
——, 'Western Art and the Perception of Space', *Storia dell'arte*, 62: (1988): 5–13.
Gonneau, Pierre, *A l'arbe de la Russie muscovite: Serge de Radonège et André Roublev: Légendes et images* (*XIVe-XVIIe siècles* (Institut d'Etudes Slaves, 2007).
Gould, Stephen, *Time's Arrow, Time's Cycle: Myth and Metaphor in the Discovery of Geological Time* (Cambridge, MA, 1987).
Grabar, André, 'Plotin et les origins de l'esthétique médiévale', *Cahiers archéologiques: Fin de l'Antiquité et Moyen Age*, 1 (1945): 15–34.
——, *Martyrium: reserches sur le culte des religues et l'art chrétien antique* (Paris, 1946).
Grabar, Oleg, *The Formation of Islamic Art*, revised and enlarged edition (New York and London, 1987).
Gray, Camilla, *The Russian Experiment in Art 1863–1922* (London, 1986).
Greenberg, Clement, 'Towards a Newer Laocoon' (1940) in his *Collected Essays and Criticism*, vol. 1 (Chicago and London, 1986), pp. 23–38.
Nazianzen, Gregory, *Faith Gives Fullness to Reason: The Five Theological Orations of Gregory Nazianzen* (Leiden, New York, 1991).
Grünbaum, Adolf, *Philosophical Problems of Space and Time*, second enlarged edn. (Dordrecht, 1973).
Gurevich, Aron, *Categories of the Medieval World* (London, Boston, 1985, first in Russian in 1972).
Guyer, Paul (ed.), *The Cambridge Companion to Kant* (Cambridge, 1992).
——, *Kant and the Claims of Knowledge* (Cambridge, 1987).
Habermas, Jürgen, *The Philosophical Discourse of Modernity: Twelve Lectures* (Cambridge, 1990, first in German in 1985).
Hacking, I., *Why Does Language Matter to Philosophy?* (Cambridge, 1975).
Hagstrum, Jean, *The Sister Arts: The Tradition of Literary Pictorialism and English Poetry from Dryden to Gray* (Chicago, 1958).
Hahn, Cynthia, '*Visio Dei*: Vision in Medieval Visuality' in Robert Nelson (ed.), *Visuality before and beyond the Renaissance* (Cambridge, 2000), pp. 169–97.
Hamburger, Jeffrey 'The Place of Theology in Medieval Art History', in Jeffrey Hamburger (ed.), *The Mind's Eye; Art and Theological Argument in the Middle Ages* (Princeton, 2006).

Hathaway, Ronald, *Hierarchy and the Definition of Order in the Letters of Pseudo-Dionysius* (The Hague, 1970).

Haynes, Deborah, *Bakhtin and the Visual Arts* (Cambridge, 1995).

Hegel, Georg, *Aesthetics, Lectures on Fine Art*, tr. by T. Knox (Oxford, 1975).

——, *Introduction to the Lectures on the History of Philosophy*, tr. by T. Knox and A. Miller (Oxford, 1985).

——, *Lectures on the Philosophy of World History: Introduction*, tr. by H. Nisbet (Cambridge, 1975).

Heidegger, Martin, *Poetry, Language and Thought* (New York, 1971).

Helm, Paul, *Eternal God: God without Time* (Oxford, 1988).

Helmholtz, Hermann von, *Helmholtz's Treatise on Physiological Optics* (Rochester, 1925, first in German in 1896), 3 vols.

Henderson, Linda, *The Fourth Dimension and Non-Euclidean Geometry in Modern Art* (Princeton, *1983*).

——, 'A New Facet of Cubism: "The Fourth Dimension" and "Non-Euclidean Geometry" Reinterpreted', *The Art Quarterly*, 34 (1971): 410–33.

Hildebrand, Adolf, 'The Problem of Form in the Fine Arts' (1893) in Harry Mallgrave and Eleftherios Ikonomou (eds), *Empathy, Form and Space: Problems in German Aesthetics 1873–1893* (Santa Monica, CA, 1994), pp. 227–81.

Hinton, Charles, 'What is the Fourth Dimension?', *Dublin University Magazine* (1880): 15–34.

Hirschkop, Ken, *Mikhail Bakhtin: An Aesthetics for Democracy* (Oxford, 1999).

Holquist, Michael and Katriona Clark, 'The Influence of Kant in the Early Works of Mikhail Bakhtin' in Joseph Strelka (ed.), *Literary Theory and Criticism: Festschrift for René Wellek* (Berne and Frankfurt, *c.*1984), pp. 299–315.

Holquist, Michael and Vadim Liapunov (eds), *Art and Aswerability: Early Philosophical Essays by M. M. Bakhtin* (Austin, TX., 1990).

Homer, *Odyssey*, XI (Harmondsworth, 1946).

——, *Iliad*, XXII (Harmondsworth, 1950).

Hughes, Lindsey, 'Inventing Andrey: Soviet and Post-Soviet Views of Andrey Rublev and His Trinity Icon', *Slavonica*, 3/2 (2003): pp. 83–91.

——, 'Monuments and Identity' in Simon Franklin and Emma Widdis (eds), *National Identity in Russian Culture* (Cambridge, 2004), pp. 171–97.

Innemée, Karel, 'Some Notes on Icons and Relics', in Christopher Moss and Katherine Kiefer (eds.), *Byzantine East and Latin West: Art-historical Studies in Honour of Kurt Weitzmann* (Princeton, 1995), pp. 519–23.

Iorga, Nicholai, *Byzance après Byzance: Continuation de l'Histoire de la vie byzantine* (Bucharest, 1935).

Ivanova, Elena V. (ed.) *Pavel Florensky i simvolisti* (Pavel Florensky and the Symbolists) (Moscow, 2004).

Iversen, Margaret, *Alois Riegl: Art History and Theory* (Cambridge, MA, and London, 1993).

Janaway, Christopher, *Self and World in Schopenhauer's Philosophy* (Oxford, 1989).

——, 'Knowledge and Tranquillity' in Dale Jacquette (ed.), *Schopenhauer, Philosophy, and the Arts* (Cambridge, 1996), pp. 39–62.

Jay, Martin, *Downcast Eyes: The Denigration of Vision in Twentieth Century French Thought* (Berkeley, London, 1993).

Jensen, Robin, *Understanding Early Christian Art* (London and New York, 2000).

John of Damascus, *On the Divine Images* (Crestwood, New York, 1980).

——, 'The Orthodox Faith' in *Fathers of the Church*, vol. 37: "St. John of Damascus: Writings" (New York, 1958).

Jonas, Hans, *The Phenomenon of Life: Toward a Philosophical Biology* (Chicago and London, 1966, reprint 1982).

Kalokyris, Konstantinos, 'Byzantine Iconography and Liturgical Time', *Eastern Churches Review*, 1 (1966–7): 359–63.

Kant, Immanuel, *Critique of Aesthetic Judgement*, tr. by J. C. Meredith (Oxford, 1911).

Karlin-Hayter, Patricia, 'Iconoclasm' in Cyril Mango (ed.), *The Oxford History of Byzantium* (Oxford, 2002), pp. 153–69.

Kartsonis, Anna, 'The Responding Icon' in Linda Safran (ed.), *Heaven on Earth: Art and Church in Byzantium* (Pennsylvania, 1998), pp. 58–81.

Kemp, Martin, 'From *Mimesis* to *Fantasia*: Quattrocento Vocabulary of Creation, Inspiration and Genius in the Visual Arts', *Viator*, 8 (1976): 347–99.

——, 'Science, Non-science and Nonsense: Interpretation of Brunelleschi's Perspectice', *Art History*, 1/2 (1978): 134–62.

——, *Leonardo da Vinci: The Marvellous Works of Nature and Man* (London, Melbourne, 1981).

——, *The Science of Art: Optical Themes in Western Art from Brunelleschi to Seurat* (London and New Haven, 1990).

——, *Behind the Picture: Art and Evidence in the Italian Renaissance* (New Haven and London, 1997).

——, *Spectacular Bodies: The Art and Science of the Human Body from Leonardo to Now* (London, Berkeley, 2000).

——, *Visualizations: The Nature Book of Art and Science* (Oxford and New York, 2000).

——, *Leonardo on Painting: An Anthology of Writings with a Selection of Documents Relating to His Career as an Artist* (New York and London, 1989, reprint 2001).

——, 'From Different Points of View: Correggio, Copernicus and the Mobile Observer' in Lars Jones and Louisa Matthew (eds.), *Coming About ...A Festschrift for John Shearman* (Cambridge, MA, 2001), pp. 207–15.

Kenny, Anthony, *The God of the Philosophers* (Oxford, 1979).

Kern, Stephen, *The Culture of Time and Space 1880–1918* (London, 1983).

Khoruzii, Sergei, 'Filosofskii simvolizm Pavla Florenskogo i ego zhiznenie iztoki' (The Philosophical Symbolism of Pavel Florensky and Its Life Sources) in Isupov, K. (ed.), *Florensky: Pro et Contra* (St. Petersburg, 1996), pp. 525–58.

Kierkeggard, Søren, *Practice in Christianity* (Princeton, 1991, first in Danish in 1850).

Kitzinger, Ernst, *The Art of Byzantium and the Medieval West* (Bloomington, London, 1976).

Klee, Paul, 'Creative Credo' (1920) in Herschel Chipp (ed.) *Theories of Modern Art: A Source Book by Artists and Critics* (Berkeley and London, 1968), pp. 182–87.

Kneale, William, 'Time and Eternity in Theology', *The Proceedings of the Aristotelian Society*, 61 (1960–61): 87–108.

Koselleck, Reinhart, *Zeitschichten. Studien zur Historik* (Frankfurt, 2000).

Krumbacher, Karl, *Geschichte der byzantinischen Literatur* (Munich, 1891).

Kuz'min, Nikolay, 'Andrey Rublev', *Novyi mir*, 10 (1960): 206.

Ladin, Jay, 'Fleshing out the Chronotope' in Caryl Emerson (ed.), *Critical Essays on Mikhail Bakhtin* (New York, 1999).

Ladner, Gerhart, 'The Concept of the Image in the Greek Fathers and the Byzantine Iconoclastic Controversy', *Dumbarton Oaks Papers*, 7 (1953): 1–34.

Lampe, Geoffrey W. H., *A Patristic Greek Lexicon* (Oxford, 1961).

Laporte, Paul, 'Cubism and Science', *Journal of Aesthetics and Art Criticism*, 7/3 (1949).

Latour, Bruno 'Opening One Eye While Closing the Other … A Note on Some Religious Paintings' in Gordon Fife and John Law (eds.), *Picturing Power: Visual Depiction and Social Relations* (London, 1988).

Lasarev, Viktor, 'La Trinité d'André Roublev', *Gazette des Beaux-Arts*, (1959) : 289–300.

——, *Andrei Rublev i ego shkola* (Moscow, 1960).

——, *Feofan Grek i ego shkola* (Theophanes the Greek and His School) (Moscow, 1961, German translation from 1968).

Leadbeater, Charles, *The Astral Plane: Its Scenery, Inhabitants and Phenomena* (London, 1895).

——, *Clairvoyance* (London, 1903, first in 1899).

Lee, Rensslelaer, '*Ut Pictura Poesis:* The Humanistic Theory of Painting', *Art Bulletin*, 22 (1940): 197–270.

Leeuw, Gerardus van der, *Sacred and Profane Beauty: The Holy in Art* (New York, Chicago, 1963, first in Dutch in 1932).

Leftow, Brian, *Time and Eternity* (Ithaca and London, 1991).

Leibniz, Gottfried, *Philosophical Essays*, eds. R. Ariew and D. Garber (Indianapolis, Cambridge, 1989).

——, *The Leibniz-Arnauld Correspondence*, ed. and tr. H. Mason (New York, 1967).

Lessing, Gotthold, *Laokoon: An Essay on the Limits of Painting and Poetry* (1766) (Baltimore, London, 1984).
Leyden, W. von, 'Time, Number and Eternity in Plato and Aristotle', *Philosophical Quarterly*, 14 (1964): 35–53.
Lipsey, Roger, *An Art of Our Own: The Spiritual in Art* (Boston, 1988).
Lloyd, G. R., 'Views on Time in Greek Thought' in Louis Gardet et al. (ed), *Cultures and Time* (Paris, 1976), pp. 117–49.
Lossky, Nikolay, *History of Russian Philosophy* (London, 1952).
Lossky, Vladimir, *The Mystical Theology of the Eastern Church* (Cambridge and London, 1957).
Louglin, Gerald, 'Time' in Adrian Hastings, Alistair Mason, and Hugh Pyper, H. (eds), *The Oxford Companion to Christian Thought* (Oxford and New York, 2000), pp. 707–709.
Louth, Andrew, 'Iconoclasm' in Adrian Hastings, Alistair Mason, and Hugh Pyper, H. (eds), *The Oxford Companion to Christian Thought* (Oxford and New York, 2000), pp. 318–20.
Lukács, Georg, *The Destruction of Reason* (London, *c.*1980).
Maeterlinck, Maurice, 'The Knowledge of the Future' in his *The Unknown Guest*, (London, 1914).
Mahoney, Edward, 'Pseudo-Dionysius's Conception of Metaphysical Hierarchy and Is Influence on Medieval Philosophy', *Recontres de philosophie médiévale*, 7/2 (2000): 429–76.
Mango, Cyril, *The Art of the Byzantine Empire 312–1453* (Englewood Cliffs, 1972).
Marsh, J., 'Christian Worship: Human and Divine Transcendence of Time' in John Booty (ed.), *The Divine Drama in History and Liturgy* (Alison Park, PA, 1984), pp. 123–31.
Martin, Edward J., *A History of the Iconoclast Controversy* (London and New York, 1930).
Mathew, Gervase, *Byzantine Aesthetics* (London, 1963).
——, 'Abstract of a Lecture on the Character of the Theological Divergence between East and West', *Eastern Churches Review*, 4–5 (1972–3): 116–18.
Melville, Stephen, 'The Temptation of New Perspectives', *October*, 52 (1990): 3–15.
Merleau-Ponty, Maurice, 'Eye and Mind' in *The Primacy of Perception* (Evanston, Ill., 1964).
——, 'Eye and Mind' in Osborne, H. (ed.), *Aesthetics* (London and Prescot, 1972), pp. 55–86.
Meyendorff, John, *Byzantine Theology: Historical Trends and Doctrinal Themes* (New York, 1974).
Migne, Jacques-Paul (ed.), *Patrologiae cursus completes. Series Graeca* (Paris, 1857–66).

Mikhailov, Alexander, 'O. Pavel Florensky kak filosof granitsi' (Father Pavel Florensky as a Philosopher of the Boundary), *Voprosy iskustvoznania* (Problems in the Theory of Art), 4 (1994).

Mikhailov, Boris, 'Coming of Age: Russian Art History as a Professional Discipline in the 1910s and 1920s', *Experiment*, 3 (1997): 7–14.

Milesk, Margaret, *The Image as Insight: Visual Understanding in Western Christianity and Secular Culture* (Boston, 1985).

Miller, Arthur, *Einstein, Picasso: Space, Time and the Beauty That Causes Havoc* (New York, 2001).

Misler, Nicoletta, 'Toward an Exact Aesthetics: P. Florensky and the Russian Academy of Artistic Sciences' in John Bowlt and Olga Matich (eds.), *Laboratory of Dreams: The Russian Avant-garde and Cultural Experiment* (Stanford, CA, 1996), pp. 118–33.

——, 'Pavel Florensky as Art Historian' in Pavel Florensky, *Beyond Vision: Essays on the Perception of Art* (London, 2002), pp. 29–93.

Mitchell, W. J. Thomas, *Iconology: Image, Text, Ideology* (Chicago and London, 1986).

Momigliano, Arnaldo, 'Time in Ancient Historiography', Suppl. 6: 'History and the Concept of Time', *History and Theory* (1966): 1–23.

Morson, Gary and Caryl Emerson, *Mikhail Bakhtin: The Creation of a Prosaic* (Stanford, 1990).

Nazianzen, Gregory, *Faith Gives Fullness to Reasoning: The Five Theological Orations of Gregory Nazianzen* (Leiden, New York, 1991).

Neave, Richard, 'Richard Neave's Egyptian Encounter' in John Prag and Richard Neave, *Making Faces: Using Forensic and Archaeological Evidence* (College Station, TX, 1997), pp. 41–53.

Neher, André, 'The View of Time and History in Jewish Culture' in Gardet (ed.), *Culture and Time* (Paris, 1976).

Neil, Bronwen, 'The Western Reaction to the Council of Nicaea II', *Journal of Theological Studies*, New Series, 51 (2000): 533–53.

Neisser, H., 'Are Time and Space Real?', *Philosophy and Phenomenological Research*, 31/3 (1971): 421–25.

Nekrasova, Elena, 'The Unrealized Project of the 1920s for the Creation of 'Symbolarium' and Its First Issue 'Point' (in Russian) in *Pamiatniki kulturi. Novie otkritia* (Monuments of Culture. New Discoveries) (Leningrad, 1994), pp. 99–115.

Newton-Smith, William, *The Structure of Time* (London, 1980).

Nietzsche, Friedrich, *The Birth of Tragedy* (1872) (Harmondsworth, 1993).

Obolensky, Dmitri, *The Byzantine Commonwealth: Eastern Europe, 500–1453* (London, 1971).

Odinokov, Feodor, 'Obretenie Andreia Rubleva', *Nauka i religiia*, 10 (1971): 86.

O'Donnell, *Trinity and Temporality* (Oxford, 1983).

Oikonomides, Nicholas, Nicholas, 'The Holy as Asset', *Dumbarton Oaks Papers*, 45 (1991): 35–45.

Olin, Margaret, M., *Forms of Representation in Alois Riegl's Theory of Art* (University Park, PA, 1992).
Osborne, Harold, *The Oxford Companion to Art* (Oxford, 1970).
Ouspensky, Leonid, *The Theology of the Icon* (Crestwood, New York, 1978).
Pächt, Otto, *The Rise of Pictorial Narrative in Twelfth Century England* (Oxford, 1962).
——, 'Art Historians and Art Critics', *The Burlington Magazine* (1963): 190–91.
——, *Practice of Art History: Reflections of Method* (London, 1999, first in German in 1986).
——, 'The End of Image Theory' (1930–1931) in Christopher Wood (ed.), *The Vienna School Reader: Politics and Art Historical Method in the 1930s* (New York, 2000), pp. 181–99.
Padgett, Alan, *God, Eternity and the Nature of Time* (New York, 1993).
Panofsky, Erwin, *Gothic Architecture and Scholasticism* (Latrobe PA, reprint 2005, first in 1951).
——, 'History of the Theory of Human Proportions as a Reflection on the History of Style' in his *Meaning in the Visual Arts* (Harmondsworth, 1970), pp. 82–139.
——, 'The Concept of Artistic Volition' (1920), *Critical Inquiry*, 8 (1981): 7–34.
——, *Perspective as a Symbolic Form* (New York, 1997).
Pares, Bernard, *A History of Russia* (New York, 1965).
Pater, Walter, *The Renaissance* (1873) (Oxford and New York, 1986).
Pausanias, *Guide to Greece* (Harmondsworth, 1971).
Pelikan, Jaroslav, *The Christian Tradition: A History of the Development of Doctrine*, vol. 2: *The Spirit of Eastern Christendom* (Chicago, 1971–89).
Perl, Eric, 'Hierarchy and Participation in Donysius the Areopagite and Greek Neoplatonism', *American Catholic Philosophical Review*, 68/1 (1994): 15–30.
Piatnitsky, Yuri et al. (eds), *Sinai, Byzantium, Russia: Orthodox Art from the Sixth to the Twentieth Century* (St. Petersburg, 2000).
Pike, Nelson, *God and Timelessness* (London, 1970).
Podro, Michael, *The Critical Historians of Art* (New Haven and London, 1982).
Popova, Olga, 'Medieval Russian Painting and Byzantium' in Roderick Grierson (ed.), *Gates of Mystery: the Art of Holy Russia* (Cambridge, 1994).
Porter, Laurin, 'Bakhtin's Chronotope: Time and Space in 'A Touch of the Poet' and 'More Stately Mansions', *Modern Drama*, 43/3. (1991): 369–82.
Proclus, *A Commentary on the First Book of Euclid's Elements*, tr. by G. Morrow (Princeton, 1970).
Procopius, *The Secret History* (Harmondsworth, 1981).
Pseudo-Dionysius, *The Complete Works* (London, 1987).
Punin, Nikolay, *Apollon*, 2 (1915):19–20.
Purishev, Boris, 'Andrey Rublev i obshchie vaprosi razvitie drevnerusskogo isskustvo 14–17 vv' (Andrey Rublev and Some General Questions of the Development of Ancient Russian Art, 14th–17th Centuries) in Boris

Mikhailovski and Boris Purishev, *Ocherki istorii drevnerusskoi monumental' noi zhivopisi* (Notes on the History of Ancient Russian Monumental Painting) (Moscow and Leningrad, 1941).

Quenot, Michel, *The Icon: Window on the Kingdom* (London, 1992, first in French in 1987).

Quincey, Thomas De, *Confessions of an Opium Eater* (1821) (Harmondsworth, 1971).

Rampley, Matthew, 'From Symbol to Allegory: Warburg's Theory of Art', *Art Bulletin*, LXXIX/11 (1997): 41–57.

Ratzinger, Joseph, *The Spirit of Liturgy* (San Francisco, 2000).

Raushenbach, Boris, *Prostranstvennie postroeniia v drevnerusskoi zhivopisy* (Spatial Constructions in Ancient Russian Painting) (Moscow, 1975).

Reichenbach, Hans, *The Direction of Time* (Berkeley, 1956).

Rescher, Nicholas, *The Philosophy of Leibniz* (Englewood Cliffs, New Jersey, 1967).

Reynolds, Joshua, *Discourses* (London, 1905).

Richardson, John A., *Modern Art and Scientific Thought* (Urbana, 1971).

Riegl, Alois, *Late Roman Art Industry* (Rome, 1985, first in German in 1901).

——, *The Group Portraiture of Holland* (Los Angeles, CA, 1999, first in German in 1903).

Ringbom, Sixten, 'Art in the "Epoch of the Great Spiritual". Occult Elements in the Early Theory of Abstract Painting', *Journal of the Warburg and Courtauld Institutes*, 29 (1966): 386–418.

——, *The Sounding Cosmos. A Study in the Spiritualism of Kandinsky and the Genesis of Abstract Painting* (Abo, Finland, 1970).

Robinson, Stuart, *Images of Byzantium* (London, 1996).

Rogozhin, Andrei, *Put' teosofii* (The Path of Theosophy) (Petrozavodsk, 1992).

Rosenthal, Bernice (ed.), *Nietzsche in Russia* (Princeton, 1986).

——, *The Occult in Russian and Soviet Culture* (Ithaca and London, 1997).

Rudavsky, Tamar, *Time Matters: Time, Creation and Cosmology in Medieval Jewish Philosophy* (Albany, 2000).

Runciman, Stephen, 'The Schism between Eastern and Western Churches', *Anglican Theological Review*, XLIV/4 (1962): 337–50.

——, *The Fall of Constantinople 1453* (Cambridge, 1963).

Russian Primary Chronicle, tr. and ed. Samuel H. Cross and Olgerd Sherbowitz-Wetzor (Cambridge, MA, 1953).

Savile, Anthony, *Routledge Philosophy Guidebook to Leibniz and the Monadology* (New York and London, 2000).

Schelling, Friedrich, *System of Transcendental Idealism* (1800) (Charlottesville, 1978).

——, *Philosophy of Art* (1859) (Minneapolis, 1989).

Schlegel, August, *Vorlesunger über schöne Literatur und Kunst*, vol. 1: *Die Kunstlehre* (Theory of Art) (Stuttgard, 1963).

Schmemann, Alexander, *Introduction to Liturgical Theology* (London and Portland, Maine, 1966, first in Russian in 1966).
——, *Russian Theology: A Bibliographical Survey* (Virginia, 1969).
Scholtz, Bernhard, 'Bakhtin's Concept of the 'Chronotope': The Kantian Connection' in David Shepherd (ed.), *The Contexts of Bakhtin: Philosophy, Authorship, Aesthetics* (Australia, Canada, 1998), pp. 141–73.
Schopenhauer, Arthur, *The World as Will and Representation* (New York, 1958, reprint 1969, first in German in 1819).
——, *Arthur Schopenhauer: Manuscript Remains in Four Volumes*, vol. 1 (Oxford, 1989–1990).
Schweizer, Nikolas, *The Ut Pictura Poesis Controversy in Eighteenth Century England and Germany* (Bern, 1972).
Sendler, Egon, *The Icon: Image of the Invisible* (Redondo Beach, CA, 1988, first in French in 1981).
Setton, Kenneth, *The Christian Attitude towards the Emperor in the Fourth Century* (New York and London, 1941).
Shearman, John, *Only Connect ... Art and the Spectator in the Italian Renaissance* (Princeton, 1992).
Siderov, Alexei, 'The Artistic Creativity of Dreams', *Experiment*, 3 (1997): 179–80.
Sidoti, Antoine, *La Prose du Transsibérien et de la Petite Jehanne de France: Blaise Cendrars, Sonia Delaunay, nov. – dec. 1912 – juin 1914 : genése et dossier d'un polemique* (Paris, 1987).
Sorabji, Richard, *Time, Creation and the Continuum* (Ithaca and London, 1983).
Sourian, Etienne, 'Time in the Plastic Arts', *Journal of Aesthetics and Art Criticism*, 7/4 (1949): 294–307.
Steinberg, Leo, 'Leonardo's *Last Supper*', *The Art Quarterly*, 36 (1973): 297–410.
Stump, Eleonore and Norman Kretzmann, 'Eternity', *The Journal of Philosophy*, 78/8 (1981): 429–58.
Summers, David, 'Representation' in Robert Nelson and Richard Shiff (eds), *Critical Terms for Art History* (Chicago and London, 1996), pp. 3–17.
Swinburne, Richard, *The Coherence of Theism* (Oxford, 1977).
Talbot, Alice-Mary (ed.), *Byzantine Defenders of Images: Eight Saints' Lives in English Translation* (Washington, DC, 1998).
Tanner, Michael, *Schopenhauer: Metaphysics and Art* (London, 1998).
Tanner, Norman (ed.), *Decrees of the Ecumenical Councils*, (London and Washington, 1990).
Tarán, Leonardo, *Parmenides: A Text with Translation, Commentary and Critical Essays* (Princeton, 1965).
Tarasov, Oleg, *Icon and Devotion: Sacred Places in Imperial Russia*, tr. and ed. R. Milner-Gullard, (London, 2002, first in Russian in 1995).
Tatarkiewicz, Wladyslaw, *History of Aesthetics*, vol. 2: *Medieval Aesthetics* (The Hague and Warsaw, 1970).

Teichmann, Roger, *The Concept of Time* (Basingstoke, 1995).

Terras, Victor, 'The Aesthetic Categories of *Ascent* and *Descent* in the Poetry of Viacheslav Ivanov', *Russian Poetics*, 1975.

Theodore the Studite, *On the Holy Icons* (Crestwood, New York, 1981).

Tihanov, Galin, *The Master and the Slave: Lukács, Bakhtin, and the Ideas of Their Time* (Oxford, 2000).

Todorov, Tzvetan, *Theories of the Symbol* (Oxford, 1982, first in French in 1977).

——, 'From *Mikhail Bakhtin: The Dialogical Principle*' in Caryl Emerson (ed.), *Critical Essays on Mikhail Bakhtin* (New York, 1999).

Trubetskoy, Evgeny, *Umozrenie v kraskakh: tri ocherka o russkoi ikone* (A Contemplation in Colours: Three Essays on the Russian Icon) (Paris, 1965, first in 1915–18, English translation from 1973).

Tscherning, Marius, *Physiological Optics, Dioptrics of the Eye, Functions of the Retina, Ocular Movements, and Binocular Vision* (Philadelphia, 1904, first in French in 1898).

Underwood, Paul, 'The Evidence Restoration in the Sanctuary Mosaics of the Dormition at Nicaea', *Dumbarton Oaks Papers*, 13 (1959): 235–43.

Uspensky, Boris, *The Semiotics of the Russian Icon* (Lisse, 1976, first in Russian in 1971).

Uspensky, Boris, A. A. Dorogov, and Viacheslav Ivanov, 'P. A. Florensky i ego statiia "Obratnaia perspektiva" (P. A.Florensky and His Essay 'Reverse Perspective', *Trudy po znakovim sistemam* (Researches on Sign Systems), Tartu, 3 (1967).

Vallier, Dora, *L'art abstrait* (Paris, 1980).

Vasilenko, L. I., 'O magii i okkul"tizme v nasledii o. Pavla Florenskogo', *Vestnik Pravoslavnogo Sviato-Tikhonovskogo Gumanitarnogo universiteta*, 3 (2004): 81–99.

Vasiliev, Aleksander, 'The Iconoclastic Edict of Caliph Yazid II', *Dumbarton Oaks Papers*, 9–10 (1955–56): 23–49.

Vikan, Gary, 'Relics and Icons' in Myrtali Acheimaston-Potamianon (ed.), *Holy Image, Holy Space*, exhibition catalogue (Athens and Baltimore, 1988), pp. 45–8.

Vipper, Boris, 'Problema vremeni v izobrazite 'lnom iskusstve' (The Problem of Time in the Visual Arts) in *50 let Gosudarstvennomu mezeiu izobrazitel'nykh iskusstv imeni A.S.Pushkin. Sbornik statei* (A 50 Year Anniversary of the State Museum of the Visual Arts "A.S. Pushkin". A Collection of Articles) (Moscow, 1962), pp. 134–50.

Vischer, Friedrich, 'Das Symbol' in Friedrich Vischer (ed.), *Philosophische Aufsätze, Eduard Zeller zu seinem 50 jahrigen Doctor-Jubiläum gewidmet* (Leipzig, 1887), pp. 153–93.

Voloshinov, Alexander, 'The Old Testament Trinity of Andrey Rublyov: Geometry and Philosophy', *Leonardo*, 32/2 (1999).

Vzdrovov, Gerol'd (ed.), *Troitsa Andreia Rubleva: Antologiia* (The Trinity of Andrey Rublev: An Anthology) (Moscow, 1981).

Ware, Kallistos, 'The Theology of the Icon: A Short Anthology', *Eastern Churches Review* (1976–7): 3–9.

——, 'The Theology and Spirituality of the Icon' in Royal Academy of Arts, *From Byzantium to El Greco: Greek Frescos and Icons* (London, 1987 exhibition), pp. 37–9.

Weyl, Hermann, *Space, Time, and Matter* (London, 1922).

Whittaker, John, *God, Being, Time: Two Studies in the Transcendental Tradition in Greek Philosophy* (Oslo, 1971).

Whitrow, Gerald J., *Time in History: Views on Time from Prehistory to the Present Day* (Oxford and New York, 1989).

Wilde, Oscar, 'The Critic as Artist' in *The Works of Oscar* (ed.) G. Maine (London and Glasgow, 1948), pp. 948–99.

Williams, Rowan, 'Russian Christian Thought' in Adrian Hastings, Alistair Mason, and Hugh Pyper (eds), *The Oxford Companion to Christian Thought* (Oxford and New York, 2000), pp. 630–33.

Wimsatt, William, 'Laokoon: An Oracle Reconsidered' in his *The Day of the Leopards: Essays in Defense of Poetry* (New Haven and London, 1976).

Wind, Edgar, 'Warburg's Concept of *Kulturwissenschaft* and Its Meaning for Aesthetics'(1931) in his *The Eloquence of Symbols: Studies in Humanistic Art* (Oxford, 1950, reprint,1983).

Winfield, June and David Winfield, *Proportion and Structure of the Human Figure in Byzantine Wall-painting and Mosaic* (Oxford, 1982).

Wolfson, Harry, *The Philosophy of Spinoza* (New York, 1969).

Wolterstorff, Nicholas, 'God Everlasting' in Clifton Orlebeke and Lewis Smedes (eds.), *God and the Good: Essays in Honour of Henry Stob* (Grand Rapids, 1975).

——, *Art in Action: Toward a Christian Aesthetic* (Michigan, 1980).

Worringer, Wilhelm, *Abstraction and Empathy: A Contribution to the Psychology of Style*, London, 1953, first in German in 1908).

Wulff, Oscar, 'Die umgekehrten Perspektive und die Niedersicht. Eine Raumanschaungsform der altbyzantischen Kunst und ihre Fortbildung in der Renaissance' in *Kunstwissenschaftliche Beiträge A. Schmarsow gewidmet* (Leipzig, 1907).

Wybrew, Hugh, *The Orthodox Liturgy: The Development of Eucharistic Theology in the Byzantine Rite* (Crestwood, New York, 1989, reprint 1990).

——, *Orthodox Feasts of Mary and Christ: Liturgical Texts with Commentary* (London, 1997).

Yiannias, John, 'A Re-examination of the 'Art Statute' in the Acts of Nicaea II', *Byzantinische Zeitschrift*, 80 (1987): 348–59.

Young, Julian, 'Schopenhauer, Heidegger, Art, and Will' in Dale Jacquette (ed.), *Schopenhauer, Philosophy, and the Arts* (Cambridge, 1996), pp. 162–81.

——, 'The Standpoint of Eternity: Schopenhauer on Art', *Kant-Studien*, 78 (1987): 424–41.

Zenkovsky, Vasilii, *A History of Russian Philosophy* (New York, 1953).

Zernov, Nicholas, *The Russian Religious Renaissance of the Twentieth Century* (London, 1963).

Zhegin, Lev, *Iazik zhivopisnogo proizvedeniia: uslovnost drevnego izkusstvo* (The Language of the Work of Art: Conventionality of Ancient Art) (Moscow, 1970).

——, 'Vozpomenaniia o P. A. Florenskom' (Memories of P. A. Florensky) in Konstantin Isupov (ed.), *P. A. Florensky: Pro et Contra* (St. Petersburg, 1996), pp. 162–73.

Index

non-religious?
whose space?

earnest. naive re
interdisciplinary study.

sloppy

p. 142. no authority supports
her hypothesis

anachronistic
appeal : Leibnitz

east + west supposedly interacted
but language - ref to
prototype, assumed Armenian
& images + relates only to
east